Primary Crushing Plant Design

by
Frank W. McQuiston, Jr.
Robert S. Shoemaker

Society of Mining Engineers
of
American Institute of Mining, Metallurgical, and Petroleum Engineers, Inc.
New York, New York
1978

Printed in the United States of America
by Port City Press, Baltimore, MD

Library of Congress Card Number 77-94869

ISBN 0-89520-252-2

PREFACE

This monograph on the design of primary crushing installations naturally could not attempt to show more than a few of the hundreds of primary crushers installed throughout the world. It does show, however, what the authors consider to be designs representative of those which have been applied on a variety of ores and under a variety of conditions.

While ores may be tested thoroughly in the laboratory and pilot plant for their response to many metallurgical processes, primary crushing plant design must remain the result of experience on hopefully similar ores since actual operating conditions on an unknown orebody can never be forecast. While some ores present no primary crushing problems, many others are wet and sticky and some break upon blasting to produce boulders so large they will not enter any crusher. Under these circumstances it is a credit to mining companies and engineers that so many fine installations have been made. With the advent of larger and larger mine haulage trucks for ore transport continuing challenges will be presented to these same people.

Frank W. McQuiston, Jr.
Robert S. Shoemaker

January 1, 1978

ACKNOWLEDGMENT

The authors wish to express their sincere appreciaton to the indispensable help which was given to them in the preparation of the drawings by Gordon Leitch, Peter Van Motman, and Russell Troy. We would also like to recognize with our thanks Miss Jacqueline Montgomery, who so very ably organized and typed the manuscript. The extensive support of the Mining and Metals Division of Bechtel Corporation made the entire project possible.

The authors also wish to express their thanks to the following people and companies who furnished drawings, operating data, and descriptions of their primary crushing installations. It is this willingness to share in technology that makes the mining profession the great one it is.

James J. Coile, Manager Process and Production Control (now with The Anaconda Company)	ANAMAX
A. D. Rovig, Manager Butte Operations	The Anaconda Company
Thomas R. Edwards, Manager Sacaton Unit	ASARCO
George Gauci Concentrator Superintendent	Bougainville Copper Company
James Sletcher, Producton Engineer	Canadian Johns-Manville Co., Ltd.
Jay D. McBeth, General Manager	Carlin Gold Mining Company
Allen Born, Mill Superintendent A. K. Smith, Chief Engineer Mark Koenig, Planning Engineer Gary Prin, Project Engineer	Climax Molybdenum Company
Everett Smith, Vice President Operations	Duval Corporation

Clair C. Chamberlain Resident Manager	Granduc Operation Company
James Lundy, Superintendent Crushing Plant	Inspiration Consolidated Copper Company
B. M. Monaghan, Vice President Operations	Iron Ore Company of Canada
S. D. Michaelson, Chief Engineer (Retired) B. A. Samuelson, Project Manager	Kennecott Copper Corporation
Charles Reno, Vice President and General Manager (now with Ralph M. Parsons Co.) J. McManus General Mill Superaintendent	Lornex Mining Corporation, Ltd.
Henry R. Eriksson	Luossavaara-Kiirunavaara AB
G. W. Dopson, Mill Superintendent F. C. Davis, Senior Metallurgist	Magma Copper Company
Warren Warhol, Assistant General Manager	Molycorp Incorporated
Merlyn G. Woodle, President A. M. Moyer, Assistant Chief Engineer	Reserve Mining Company
George Beals, General Manager (now with Amoco Minerals)	Palabora Mining Company, Ltd.
W. G. Martin, General Manager	Similkameen Mining Company, Ltd.
R. H. Hunt, General Manager (now with ASARCO) Leonard Dejong, Chief Metallurgist	Southern Peru Copper Company
Wayne Gould, Metallurgical Superintendent (Retired)	Cities Service Company

42" GYRATORY
54" GYRATORIES
60" GYRATORIES
72" GYRATORY
JAW CRUSHERS
UNDERGROUND CRUSHERS
REFERENCES

TABLE OF CONTENTS

PRIMARY CRUSHING

History

The earliest U. S. patent on a crushing machine was issued in 1830. The device incorporated the drop hammer principle later used in the famous stamp mill, whose history is so intimately linked with that of the golden age of American Mining. Eli Blake invented the first successful mechanical rock-breaker--the Blake jaw crusher--patented in 1858. The idea of his mechanical principle, the powerful toggle linkage, was so sound that today the Blake-type jaw crusher is the standard by which all jaw crushers are judged. In 1881, Philetus Gates was granted a patent on a machine which included in its design all the essential features of the modern gyratory crusher.

For some years after these pioneer machines were developed, crushing requirements were quite simple. All underground mining and open-pit quarrying was done by hand with tonnages and rock sizes generally small.

In the milling of precious metal ores, stamp mills were popular as the final reduction machine. The stamps were fed with an ore size readily produced by one break through small gyratory or jaw crushers. Even in large underground mining operations, there was no demand for large crushers; and increasing tonnage requirements were met by duplicating the small units. The Homestake gold mine in South Dakota was a good example, where in 1915 twenty-two small Gates gyratory crushers were used to prepare the ore for the batteries of 2,500 stamps.

Open pit mines and the introduction of the steam shovel created a demand for larger crushers, which led to the development of a gyratory with an 18 in. (457 mm) receiving opening. Up to this time, the jaw crusher had kept pace with the gyratory, both from the standpoint of receiving opening and capacity; but now the gyratory stepped into a 15-year lead with larger and larger sizes. This trend toward large primary crushers started just before the turn of the century, and in 1910 crushers with 48 in. (1.22 m) receiving openings were being built. About this time, the jaw crusher was revived in a great contribution to primary crushing by the introduction of the 84 in. by 60 in. (2.13 by 1.52 m) machine. This crusher was usually followed by a 24 in. (610 mm) opening gyratory crusher for secondary breakage, a system which reawakened the industry to the use of the jaw crusher as a primary breaker.

It is interesting that although his machines never came into general use, Thomas A. Edison ranks as a pioneer in the development of the large primary breaker. Edison reasoned that the recoverable electrical energy in a pound of coal was approximately equal to the available energy in one pound of 50% dynamite. The cost of a pound of coal, however, was about 1/100 of the cost per pound of dynamite, large amounts of which were consumed in secondary breaking to reduce ore to sizes which could be handled by the then available small primary crushers. The obvious conclusion was that it would be much cheaper to break the larger pieces of ore by mechanical rather than explosive energy. He set out to design a larger primary breaker, a development culminating in the huge, 8 ft by 7 ft (2.44 m by 2.13 m) Edison rolls. Although several were installed, theycould never compete with the more versatile gyratory and jaw types.

During these same years, when the mining industry was concerned with the development of larger and larger primaries, another type of crusher was introduced: the single, sledging roll crusher. In 1911, a 36 in. (914 mm) diameter by 60 in. (1.52 m) face single roll crusher was built. The machine achieved some degree of popularity, although its field of application was limited for primary crushing service. The largest size built was a 60 in. by 84 in. (1.52 m by 2.13 m) machine.

During the last fifteen to twenty years, the large sledging roll crusher of the Hazemag type has re-emerged in the crushing of the softer iron ores of Europe, limestones, and other non-metallics. Due to its positive discharge characteristic, the single roll crusher ranks as a top performer in the handling of wet and sticky material and will crush rock containing a substantial admixture of clay.

Of the above kinds of crushers, the gyratory is most widely used today in large open pit mining operations where the use of large shovels and trucks require a large crusher opening and high capacity. For this purpose, the 54 in. (1.37 m) is the popular size; however, occasionally a 60 in. (1.52 m) gyratory is used. The first two figures of the crusher size, 54 in. by 74 in. for example, denote the width of the receiving opening; the second two figures, the maximum diameter of the crushing mantle. The most common 60 in. (1.52 m) unit has a mantle diameter of 89 in. (2.26 m); but a few 102 in. (2.59 m) and 109 in. (2.77 m) units are in service. One 72 in. by 93 in. (1.83 by 2.44 m) unit is operating on an extremely blocky asbestos ore in Canada. The jaw crusher is often selected for medium-size pit crushing operations and is usually used for underground crushing requirements. A 42 by 48 in. (1.07 by 1.22 m) or 48 by 60 in. (1.22 m by 1.52 m) are favored sizes for underground crushing.

Selection of Primary Crushers

The term "primary crusher" by common usage is applied to the crusher which takes up the job of reduction where the blasting operation leaves off.

Selecting the right type and size of primary crusher is of prime importance, as oftentimes this machine is the largest and most expensive single item of equipment in the plant. The following factors all have bearing upon the choice of primary crusher:

1. Characteristics of the material
2. Methods employed in drilling and blasting
3. Type and size of open pit equipment to be used
4. Size of mine-run ore
5. Method of feeding the crusher

Until the last few years and the advent of concentrators which were initially built with large capacities, such as Sierrita and Bougainville, instead of being expanded in stages to large capacities, as Pima and many others, primary crushers were generally selected for the size of their receiving opening. This opening was generally large enough to accept rocks loaded by five- to ten-yard shovels and transported by the 35- to 80-ton trucks then in use. The capacity which went with this opening was also generally more than sufficient to supply the concentrator, and the occasional oversize boulder could either be turned by a rock hook or removed by hoisting it out of the dump pocket without seriously delaying the mining operation. Since there are few concentrators which have not been expanded at some time, and since both shovels and trucks have been greatly increased in size, many of the 54 in. crushers which became a standard of the industry are now being pushed to their maximum capacities.

Designers of new crushing plants are having to make substantial innovations and adopt new practices in order to obtain adequate capacity with 54 in. machines, rather than purchasing the much more expensive 60 in. machines. This is being made even more the practical solution by the rapidly increasing costs of machinery, bulk materials, construction labor, and, in addition, the increasingly strong regulations on civil-structural design and safety.

Use of Grizzlies

In earlier years, with jaw crushers and smaller gyratories, it was almost standard practice to feed these crushers across grizzlies, since removal of fines from the feed was often necessary to prevent choking of the relatively small crusher settings and to increase capacities by by-passing the fines around the crushers. Even after 54 in. and 60 in. crushers were in general use, operators and designers continued to insist on grizzlies in many installations.

Strangely enough, it was the iron ore producers who began to break this pattern at the time when taconite plants began to be installed. Taconites are not sticky and in general break with a minimum of fines. Consequently, the conventional grizzly was successfully eliminated. Copper producers, with their softer and sometimes sticky ores, were slow to change; but since the early 1960s, many of the primary crushers in copper plants have not used grizzlies. Certainly there will always be ores which require a grizzly ahead of a primary crusher and the "grizzly or no grizzly" argument will never be completely settled; but it seems logical that a crushing plant should be built to crush as much ore as possible instead of screening it. In addition, if particles of less than the crusher discharge setting are left with the crusher feed, instead of bypassing the crusher through a grizzly, many of them will be crushed rock-on-rock in the primary instead of rock-on-steel in the secondary and tertiary crushers. This, of course, is a difficult theory to prove by dollar savings, since there have never been two crushers, one with a grizzly and one without, operating on the same ore.

There is one convincing example of interparticle crushing, however, at Anaconda's Butte crushers, which is described in detail in the open pit section of this monograph. A change from 7 in. to 4 in. (178 to 102 mm) spacing of the bars of the grizzly ahead of the 60 in. Butte primary crusher resulted in the three 7 ft (2.13 m) secondary crushers being able to maintain their setting when handling harder ore. Consequently, after this change rod millfeed rates did not need to be reduced nearly as much when hard ore was encountered.

Grizzlies ahead of primaries add substantially to the cost of the crusher installation and deepen the concrete enclosure by 15 to 25 ft (4.6 to 7.6 m) This height must in turn be regained by the crushed ore conveyor, which at a 15° slope must be 57 ft to 95 ft (17.4 to 29 m) longer. While a grizzly probably results in less wear on the crusher concaves and mantle, it also requires substantial maintenance. One additional disadvantage to the grizzly, which admittedly cannot be calculated, is the extra impact on the crusher by the ore falling farther from the truck. It must therefore be concluded that the installation of a grizzly ahead of a primary crusher should be made only after all factors of ore characteristics and mill operation are very carefully considered and evaluated in regard to capital, operating, and maintenance costs. In the great majority of cases, it will be found that a grizzly is not necessary. The one exception is the advantage of removing wet fines with a grizzly, which can cause acute packing in the gyratory crusher.

Design of the Dump Pocket

After the grizzly, an important consideration in the primary crusher design is the dump pocket. Dump pocket design can have a substantial effect on the cost of mining, since trucks queued up at the crusher are not productive. As a practical matter---and a practice used by most mining companies---after a truck dumps, its load should be allowed to almost clear the pocket before the next truck dumps. Otherwise, an oversize rock bridging the crusher feed opening may be buried under another full truckload of ore. Of course, if very little or no oversize is present, this problem may not occur. It would follow, then, that the dump pocket would need a capacity at each dumping position of no more than slightly over the loaded capacity of the largest truck contemplated. In practice, this would mean a dump pocket depth of no more than about 14 ft (4.27 m) is required. The width of the pocket should be wide enough to accomodate this same truck size and also be sufficiently wider to permit the rock to build up and protect the concrete walls. The Pinto Valley dump pocket is an excellent example where no liner plates are used. Only the edge of the pocket is protected from falling rock by steel.

Dump pocket dust control in mild climates is best accomplished with high-pressure water sprays located at each dumping position. Spray headers and nozzles should be in a deep groove in the dump pocket wall to be protected from falling rock. In colder climates, dust from the dump pocket is generally collected in a bag house.

A splitter in the dump pocket to protect the crusher spider is another controversial item. While some crusher manufacturers would prefer that large rocks not be allowed to hit the spider, there is no guarantee that with a splitter no rocks will fall on it. Additionally, large pieces deflected by the splitter to the sides of the pocket will often rebound and hit the spider from the side--its weakest point. A splitter also requires that the dump pocket be somewhat deeper than it would be without the splitter, thus adding substantial cost. Among 54 in. crushers without splitters are those at Similkameen and ASARCO's Mission plant. Some 60 in. crushers without splitters are Cities Service's Pinto Valley, Phelps Dodge's Tyrone, and Duval's Sierrita units.

In general, a shallow dump pocket (which is preferable) of the Similkameen or Pinto Valley type needs no splitter. Deeper dump pockets or those with grizzlies are generally equipped with splitters.

Overhead Cranes

Most primary gyratory crusher installations are equipped with overhead bridge cranes for removing the mainshaft and other hoisting duties involved in maintenance. Often, the crane is used to operate the rock hook for turning over rocks which bridge the crusher opening or hoisting oversize boulders out of the dump pocket. During the last few years, hydraulic grapples have been used extensively for clearing the crusher opening or alternatively a hydraulic impact breaker mounted on the grapple arm has been used for breaking oversize rocks. Air-operated breakers, as a rule, have not been as successful as hydraulic breakers. Grapples have been limited in capacities of 25,000 to 30,000 pounds and thus have not been too successful. One manufacturer, however, now makes a grapple of 57,000 pounds capacity which should perform well even on the largest rocks.

A few companies have now successfully installed 54 in. primary crushers without overhead cranes. Examples are Similkameen, Inspiration, Twin Buttes, and some of the cement companies. Similkameen uses a 25,000 pound grapple for removing oversize from the dump pocket, but at Inspiration a mobile crane is used to remove the occasional boulder. Mainshaft removal at Similkameen and Inspiration is done with 82- and 75-ton mobile cranes respectively. Capital cost savings by not using an overhead bridge crane are on the order of $400,000 to $500,000 at 1977 prices.

Primary Crusher Drives and Mountings

While jaw crushers must be operated by belt drives, gyratories can be belt-driven or directly connected to a slow-speed motor by a jack shaft. The latter method is generally preferred because of its simplicity, and the small extra cost of the slow-speed motor does not warrant the consideration of a belt drive.

Mounting of the crusher is best done directly on a concrete diaphragm floor with the motor on a step below the crusher. The alternate method of mounting the crusher is on a pedestal with the motor on the floor. This latter method adds to the overall height of the installation and offers no compensating advantage.

As an aid in maintenance, Palabora has installed a circular monorail around the crusher. From a trolley on the monorail is suspended an air-operated, counterweighted impact wrench which speeds up the removal or tightening of the upper shell ring bolts. Monorails should also always be installed for the removal of the jackshaft or motor if the latter cannot be reached by the crane hook.

Crushed Ore Pockets

Crushed ore pockets under gyratory crushers generally have their upper walls lined with steel to prevent wear on the concrete, while many pockets are lined with cast manganese steel plates which require much complicated imbedded metal for anchoring. Impact on these liners by crushed ore is usually light, and large sheets of AR plate make for a more economical and just as good installation as the smaller plates of cast manganese. Eye bolts installed in the ceiling of the crushed ore pocket near the walls are very useful in installing these liner plates.

The capacity of the crushed ore pocket should follow that of the dump pocket and should be not more than $1\frac{1}{2}$ times that of the largest truck contemplated. More capacity than this only adds extra expense in deepening the entire crusher structure. Level controls in the pocket are generally of the radioactive type and the source can best be suspended in a pipe cast into the concrete wall of the pocket. In the Palabora Figure 8 crusher, the radioactive source is suspended in a steel-protected enclosure in the back of the pocket with the detector installed outside the front wall. Sonic detectors for measuring high levels are much improved and are frequently used.

To aid in cleanup around the crusher itself, a hole covered with a steel plate can be installed in the ceiling of the crushed ore pocket. Spills then need not be hoisted out of the enclosure by the crane.

The entrance door to the crushed ore pocket can take many forms. At Lornex, it is hinged at the bottom, has eccentric cart rails mounted on its outer face, and is let down by a winch into the pocket to rest on two steel-protected concrete walls. At Bougainville, where there are two crushers, each crushed ore pocket has two doors on opposite sides which let down to form a bridge for the eccentric cart under the crusher. In other cases, the cart rails are permanently in place, as at Butte, Cuajone, and the Climax Henderson operation. At these two latter crushers, the eccentric carts are underslung from the wheel carriages and roll into the crushed ore pocket on permanent rails. In most instances, the rails are stored outside the pocket and are installed in the crushed ore pocket only when the eccentric is being removed.

At most crushers, the eccentric is lowered to the cart by the crane above the crusher; but in some of the newer installations, as at Palabora for example, the cart is equipped with a hydraulic platform which raises up under the eccentric and then lowers it for removal. Handling of the eccentric in this manner is easier, faster, and safer; and it frees the crane for other work.

Crushed Ore Feeders

Until recently, the traditional feeder under the crushed ore pocket was an apron or pan feeder using cast manganese pans or flights. Usually, these units operate in the range of 30 to 40 fpm (9 to 12 mpm), although the newer types using tractor parts and sealed, lubricated pins can be operated at higher speeds.

A very few primary crushers have used mechanically vibrated feeders under their crushed ore pockets (Butte and Twin Buttes, both Anaconda-designed installations). At least one company (to the authors' knowledge) has used an electromagnetic vibrating feeder under the crushed ore pocket; but this unit and the ones at Twin Buttes were subject, at times, to quite severe mechanical problems.

The first belt feeder (it is believed) to be installed under a primary crusher was the 6 ft (1.83 m) unit at Ray. This feeder, which is 75 ft (22.8 m) long, uses a V-shaped, mechanical splice; and under conditions where much tramp iron and timber is encountered, it has an average belt life of about 10 million tons. The 8 ft (2.44 m) wide belt feeder at Similkameen is vulcanized at the splice and does not encounter much tramp material. The original Similkameen belt is still in service, after handling 30 million tons, and apparently could handle as much more tonnage. It is said this belt feeder has incurred no mechanical down time. The few minor cuts which have occurred in the belt have not penetrated the carcass and have been easily repaired with cold setting rubber. The new Palabora Figure 8 crusher also uses an 8 ft (2.44 m) wide belt feeder. It is identical to the Similkameen unit, except its idlers are flat where as the former uses picking idlers so that the belt is turned upwards for a few inches at each edge. It is reported that another 8 ft wide belt feeder which was used at the Gibraltar Mine in British Columbia had a cable belt construction instead of a woven carcass. Several of these belts tore from end to end, and a fiber carcass belt which was then purchased has given satisfactory service.

The belt feeder has a number of advantages over an apron or pan feeder. It costs less, requires less headroom, exhibits radically lower maintenance costs, and can be operated faster, thus giving a more even takeaway belt loading and a higher tonnage capacity than the widest apron feeder. This last point can be specially useful as high tonnage in-pit crushing and conveying systems are installed in the larger mines. With the success that has attended the operation of the four belt feeders described and now installed, the authors believe the day of the apron feeder under primary crushers and coarse ore stockpiles is now over; and only under special circumstances should they be used in the future.

Crusher Enclosures

Gyratory crusher enclosures have traditionally been made up of square or rectangular structures of two or sometimes three compartments in plan view, with the three-compartment units having electrical and dust collection equipment in the third compartment. The two-compartment structures are less expensive and more functional because electrical gear is above ground near the maintenance area and the dust collector is similarly situated or is placed at the daylight end of the takeaway conveyor. The former has to draw dust upward but the latter draws it downward and out the tunnel. Dust generated in the dump pocket is best suppressed by water sprays located under the truck dumping position so the ore falls through a water fog curtain.

A few circular enclosures have been constructed; but, while they are economic in regard to concrete and reinforcing steel usage, support of the crusher on the resulting large-diameter diaphragm is structurally difficult and usually results in awkward support walls which limit maintenance access. The Figure-8 design of the new Palabora crusher, however, has combined low concrete and steel usage with good structural design. The enclosure for this 54 in. unit is made up of two circular, connecting silos with the smaller cylinder housing the crusher and the larger being used for maintenance, hoisting, etc.

Jaw Crushers and Impactors

The two available designs of jaw crushers are the Blake, which uses two toggles and a pitman to actuate them, and the swing jaw and the overhead eccentric or single toggle type. In the United States, the latter traditionally has been deemed a lighter duty machine while the Blake type has been used for heavy duty in the larger sizes. In Europe, single toggle units are used for heavy duty in the larger sizes, and the single toggle crusher is often choke fed, since the jaw movement tends to make it self-feeding (this is especially true in the underground mines of Sweden). In North America, at least, jaw crushers are almost always fed by an apron feeder or sometimes by a chain feeder. Some companies have used mechanical vibrating feeders because of their low cost (rather than the electromagnetic type) to feed jaw crushers; but this has often proved to be a poor choice.

Ore mined underground is almost always wet; this is often true of open pit ore as well. Wet fines, and especially those containing clays, will invariably pack on the feeder pan to form a dead load of such weight that bearings and drives will be destroyed. The dead load will also slow down the vibration and the feeding rate will decrease radically. An apron feeder is the only satisfactory method of feeding sticky ore to any crusher, jaw or gyratory, and it must be equipped with a dribble belt or the crushed ore belt must be extended under the feeder to catch fines as they fall from the return strand of the feeder.

Jaw crushers can generally handle more sticky ore than can gyratories, since they are somewhat less susceptible to packing in the crushing chamber. With sticky ores, jaw crushers are often fed over fixed or preferably vibrating grizzlies. While jaw crushers need a dump pocket, they need no crushed ore pocket and crushed ore feeder; and the overall height of a jaw crusher installation is less than that of a gyratory. Capacities, however, are limited to 1,000 to 1,300 tph of 100 pound per cubic foot (1,600 kg/m^3) ore.

While impactors and their installations are not described in this monograph, they should be noted in any discussion on the crushing of sticky ores. These units can be fed extremely large rock and are used for primary crushing where the ore is not so hard and abrasive as to require crushing in a jaw or gyratory. The low headroom required by impactors makes them suitable for use as mobile units in open pits, and they are used extensively in that manner in Europe. Very short shaft gyratories are also used in the same service. Although capacities of impactors are not in the range of large gyratories, they are effective on ores such as limestone, soft iron ores, and non-metallics.

CRUSHING IN THE PIT

Open pits and quarries are the major sources of all hard rock tonnages mined today. Normally, ore is fractured from the pit bench face by blasting and then truck-hauled to a primary crusher on the pit rim for conveying to the concentrator; overburden (or waste) is generally truck-hauled from the pit to dump areas. This system of truck haulage of blasted ore and waste accounts for some 25 to 50% of total mining costs in a typical open pit operation and commonly is the largest single cost factor. In an effort to reduce these costs mining companies are continually replacing their older and smaller trucks with larger ones and at the same time are studying conversions to conveyor haulage, since this method of transport seems to offer possibilities of lower costs per ton of material handled. In these studies it seems logical to consider installing these conveyors as close as possible to the mining face for maximum use of conveyors and minimum use of trucks, since a conveyor is much less labor intensive than a trucking system. However, direct loading onto a conveyor near the retreating bench face is impractical because of excessive lump size. Consequently, it is conventional practice to load the blasted rock into trucks at the face for haulage to the crusher. After crushing, the rock is fed onto a conveyor. The closer the crusher is to the mine, or if it is in the mine, the shorter the truck haul and the greater the potential for cost saving. At first appearance, in-pit conveying might seem to be an obvious solution, yet many mining companies have considered such a move while few have implemented it. To determine why, it is necessary to consider the factors influencing such a decision.

A conveyor system tends to be capital intensive but can have a lower operating cost per ton than trucks. Trucks, on the other hand, tend to cost less to buy but more to operate; thus, economics tend to favor the conveyor if it can be operated close to capacity and if a sufficiently large tonnage can be passed over the system to mitigate the effect of its high capital cost.

In a typical hard-rock, open-pit mine it is necessary for a variety of reasons to mine several faces of both ore and waste at the same time; and any mining system must have sufficient flexibility to permit loading from one face or another at short notice. This makes it impractical to place small mobile crushers and their consequent network of conveyors at each working face. To date no alternative has been found to using trucks for bringing all rock of the same class to some convenient marshalling point from which it may be crushed and conveyed from the pit. Another advantage of trucking when compared to conveying is flexibility. Many mines wish to deliver their shot-rock to different locations, depending on the mineral content. For instance, a typical copper mine may have to simultaneously mine milling ore, leach ore, and waste; and Palabora Mining Company in South Africa mines waste and three types of ore which must be separately crushed and milled. Such a mining operation may wish to retain the flexibility of truck haulage to change from one rock type to another or from one destination to another at a moment's notice.

Rock type and its final destination are therefore perhaps the most important criteria in determining the relative cost advantage of a conveying system. In almost all cases ore or waste, after it has been mined and loaded into trucks at the face, must be crushed before undergoing any further handling. This means the cost of conveying after crushing can be compared directly to that of trucking to determine the more economic alternative. For ore the cost of crushing can be ignored since it will eventually be required. Waste rock, on the other hand, is merely dumped and need only be crushed if this is required for conveyor handling purposes; thus, any saving gained from conveying waste instead of trucking must also overcome the cost of waste crushing.

In the case of ore handling, the question is not whether to crush but where to crush, and the plant feed conveyor may be a short connecting conveyor of a few hundred feet or a major overland conveying link several miles long. At this point the crusher location becomes very important both from the point of view of convenience and of permanence. A high percentage of the cost of a crushing installation is in concrete, labor, and reinforcing steel, none of which can be reclaimed if the crusher is moved. The problem in a deep open-pit mine is to find a location convenient to the ore faces but without shielding ore which might otherwise be mined. A mine with one absolutely permanent pit wall is a rarity.

If a future relocation is unavoidable, at least this relocation should be deferred as long as possible. The objective, of course, is maximum return on the initial crusher investment and minimum cost of any subsequent moves.

Then again, the problem of locating a crusher within the pit should not be underestimated. In addition to the obvious impact of an in-pit crusher and conveyor system on mine planning, there is the shape of the pit itself. Sound mine planning includes consideration of slope stability and general rock mechanics to reduce the dangers of slope failure. One of the important aspects of pit design includes avoidance of sharp corners or change of direction. A pit that is circular in plan is ideal and either pre-stripping to create a notch in which a crusher/conveyor can be installed or leaving a ridge on top of which one is installed will tend to introduce points of stress which could lead to eventual slope failure.

Another factor to be considered is energy consumption. Trucks are dependent upon a petroleum product whereas conveyors are run by electricity which is not necessarily dependent upon oil. Some mines have a ready access to inexpensive electric power which could be used for conveyors but are forced to use expensive fuel to run their trucks. (On the other hand, electric power costs several cents per kilowatt in some countries today.)

Against this backdrop of advantages accruing to a conveyor system must be weighted the consequent loss of flexibility. This is not only in the movement of material, already mentioned, but also as it affects downtime and the ability to change the mining plan. Obviously when a conveyor shuts down, production is completely disrupted, whereas the impact of several trucks breaking down at the same time is much less dramatic. Once a crusher and its connecting conveyor system have been installed, it is both costly and time consuming to relocate or change. Trucks, however, have a relatively short working life, usually less than ten years. A truck fleet can be replaced by similar units, units of a different size, or, if conditions warrant, a change to a completely different system can be made at that time.

The argument for in-pit crushing and conveying of waste versus trucking is exactly the same as for ore, except that the savings derived from conveying must also overcome the cost of crushing. As with crushing ore, factors other than economy must be considered. Many of the world's larger open pit mines have expanded to the point that dumping space adjacent to the mine is no longer available. This often results in a long haul beyond the pit to remote waste dumps, and there are limits to the length of haul a large off-highway rear-dump truck can make. Even if the trucks are technically capable of handling such a haul, the number of trucks required might become unwieldy, indicating the desirability of changing to another system.

There are two well-known in-pit crushing systems in the United States which bear mentioning. At ANAMAX, two 54 in. by 74 in. crushers are located so that they may crush either ore, leach rock, or waste which is then conveyed to the pit rim. At that point a transfer system can route ore or leach rock from either crusher/conveyor to the concentrator or leach plant or waste from either crusher/conveyor to the waste dump.

At the Duval Sierrita operation two crushers were originally installed near the pit rim to crush ore for transfer to the concentrator by an overland conveyor system. In later years as the pit deepened, the two original crushers were converted to handle waste by the addition of a waste conveying and stacking system. Two more crushers were then installed in the pit to crush ore which was then conveyed up the pit wall and transferred to the original overland conveyor to the concentrator.

In summary, it can be stated that in-pit crushing and conveying, particularly of ore, can offer significant advantages. However, each situation must be studied on its own merits; and no two cases are sufficiently similar to permit any all-inclusive statement of policy. The one thing that seems clear is that for a small production tonnage, a short mine life, a close plant, or a close waste dump, there seems to be little advantage in contemplating in-pit crushing and conveying at this time.

MOBILE PRIMARY CRUSHERS

Theoretically mobile primary crushers at the mine working face can be considered the ideal comminution system. A crusher which follows the ore can reduce materials handling costs because it eliminates about one ton of truck deadweight per ton of ore hauled to the crusher. It also eliminates the backhaul of this same deadweight ton.

Development of mobile primary crushers has centered in Germany where many soft iron ore and limestone deposits are flat-lying and thus readily adaptable to the use of shiftable conveyors which easily followed the working face. These ores were easily crushed in impact-type mills which had extremely large receiving openings and therefore large reduction ratios. The feeding of such a unit was done with a front-end loader which dumped its load to an inclined apron feeder, which in turn fed the impactor. Under the crusher was an inclined conveyor which transferred the crushed ore to a belt wagon for delivery to the shiftable overland conveyor. Low headroom overhead eccentric single-toggle jaw crushers and short-shaft gyratories were also used on these units, which generally travelled on tracks.

While these early mobile primary crushers were versatile and served their purpose, they generally suffered from low capacities. Manufacturers have quickly responded with larger crusher, feeders, and moveable supports, however; and today mobile crushers have capacities of up to 1500 metric tons per hour, a figure comparable to the capacity of some 54 in. by 74 in. gyratories.

An example of a quarry portable crushing plant is at the Bath cement plant of Canada Cement Lafarge, Limited. Two 12 cu yd front-end loaders are digging the stone and hauling it to the portable crushing plant which feeds a field conveyor system. The crushing plant is moved and the field conveyors extended whenever the length of the haul exceeds 300 ft. At the crushing plant, the loaders discharge the stone onto a pan feeder which feeds into the hopper of a Weserhutte-Hazemag compound crusher which combines primary and secondary rotors in the same housing. These are driven at 1200 rpm by 800 hp and 1,000 hp motors (600 and 750 kw). The maximum 3 ft feed (914 mm) is reduced to 3/4 in. (19 mm) at 800 tph. The Weserhutte steered walking mechanism, operated by hydraulic cylinders, takes five-foot strides.

Another example is at Alcoa of Australia's operation at Huntley in Western Australia. At the mine face, front-end loaders load 50 mt rear-dump trucks that transport the bauxite ore to the mobile crusher which follows the mine operation. The crusher can turn 360° within its own length, walk at a speed of 34.5 ft per minute (10.5 m per minute) and negotiate grades of 1:10. The overall weight of the unit is 560 mt, and it can crush 59 in. (1500 mm), friable bauxite rock at a rate of 1500 mtph.

Formidable specifications must be met for manufacture of a practical and economic portable crusher, and no large capacity readily portable unit suitable for large shovel operation on hard, abrasive rock has been built until recently. In 1975 a portable crushing plant using an Allis Chalmers 54 in. by 74 in. gyratory crusher was installed at Cementos Apasco S.A. in Mexico. This Weserhutte unit contains an 81 ft long by 87 in. wide (24.6 m by 2.2 m) apron feeder sloped at 25° to feed the crusher and another 25 ft long by 102 in. wide (8.25 m by 2.6 m) apron feeder which handles the 8 in. (203 mm) product. The unit is fed by two 12 cu yd (9.17 m^3) front-end loaders and operates one shift per day, six days per week. Moving the unit is accomplished by three hydraulically operated feet which walk at a speed of 4 ft per minute (1.2 m per minute). In the two and a half years it has been operated, it has been moved five times. The crusher is a standard unit; and because the crane is used for removing large rocks and cannot lift the main shaft, the crusher is equipped with a Hydroset unit for adjusting the crusher setting. Overall weight of the machine is about 700 short tons (635 mt). Normal operating tonnage at this operation is 1,000 stph (907 mtph).

MAGMA COPPER COMPANY
SAN MANUEL DIVISION
SAN MANUEL, ARIZONA

The San Manuel district of Arizona was first prospected prior to the Civil War, but there was little or no production until 1881. Small operators tried to develop ore bodies from time to time after that without success. Then, in 1917, a drilling program was carried out; but it did not indicate a copper content of interest. Finally, in the 1940s, the U. S. Geological Survey drilled the property and obtained indications of a mineralized area. At that time, Magma Copper and its principal owner, Newmont Mining Corporation, drilled many deep holes through the gravels and conglomerates and in 1948 proved an ore reserve of 367 million tons of 0.785% copper sulfide ore. Underground exploration and development were started in early 1948, and in 1953 design and construction of the surface facilities commenced. By the end of 1955 plant construction was completed.

The ore deposit of disseminated pyrite, chalcopyrite, chalcocite, molybdenite, silver, and gold is in quartz monzonite porphyry gangue. Mine production, which is entirely underground, is by full-gravity block-caving. Initially, ore mined at a rate of 36,000 tpd was hoisted through two identical shafts. From the 5,000 ton coarse ore bins located at the collar of each shaft, the ore was hauled by a shuttle service railroad in 100-ton bottom-dump rail cars a distance of seven miles to the rail bin.

Fragmentation from block-caving supplemented by underground grizzlies at the loading stations controlled the oversize material to give a feed size acceptable to the two 7 ft standard Symons crushers. During operation, however, it became evident that a large amount of secondary breaking in the grizzlies was necessary to produce a suitable feed for the standard crushers. The decision was then made to install primary gyratory crushers on the surface adjacent to the shaft site ore bins. Under each 5,000-ton coarse ore bin, an 8 ft by 20 ft (2.44 m by 6.1 m) apron feeder transfers the ore to a 13 ft 6 in. by 10 ft (4.1 m by 3.05 m) grizzly set at 6 in. (152 mm) and sloped at 45 degrees. About 65% of the ore passes the grizzly and thus bypasses the 42 in. by 65 in. (1.07 m by 1.65 m) Allis-Chalmers gyratory equipped with a Hydroset mechanism and having an open side setting of 5½ in. (140 mm). Grizzly undersize and crushed ore drop onto a 6 ft by 30 ft (1.83 m by 9.14 m) variable-speed apron feeder for transfer to a 54 in. by 90 ft 5 in. (1.37 m by 27.6 m) inclined conveyor equipped with a tramp iron detector.

Indication of tramp iron automatically stops the conveyor and both apron feeders. When the iron is removed from the belt by the operator, the feeders and conveyor are restarted. This first conveyor, sloped at 25%, discharges to a second, inclined, 48 in. by 162 ft 5 in. (1.22 m by 49.5 m) conveyor sloped at 28% from which wood is picked on the fly. A single, inclined conveyor for both iron and wood picking was not practical because stopping such a long, loaded belt (sometimes more than a thousand times a month) would have been too severe a duty for the motor and drive. Even so, the tramp iron picking conveyor is equipped with a fluid drive. After the wood picking belt, the ore is transferred to a cross conveyor and then to a railcar load-out silo which has a capacity of 12,000 st. The crushed ore silo is equipped with specially designed chutes and gates for fast loading of the railroad cars, which haul the ore to the 12,000 ton ore bin ahead of the standard and shorthead crushing plant.

The four crushers at San Manuel are operated 21 shifts per week and are down for maintenance one shift every two weeks. The present concentrator capacity of 62,000 tpd, double the size of the original plant, requires hoisting from underground through four shafts. Each shaft is equipped with its own primary unit, one of which is shown in the drawing. The production for a milling rate of 62,000 tons of ore per day makes San Manuel the largest underground operation in the world.

San Manuel's crushing cost of about $0.058 per ton is a remarkable achievement for the operation of four separate crushers, auxiliary feeders, and conveyors, and is a direct result of a good plant design and an efficient operation.

MAGMA COPPER COMPANY
SAN MANUEL DIVISION
San Manuel, Arizona

Ore type	Copper porphyry - underground mine 167 #/cu ft in place (2675 kg/cu m)
Concentrator capacity	62,000 stpd
Primary crusher	42 in. by 65 in. Allis Chalmers gyratory with Hydroset (4 units)
Connected and consumed power	300 hp - 200 hp
Capacity of coarse ore feed bin	5000 st
Crusher feeder	8 ft by 20 ft (2.44 m by 6.1 m)
Grizzlies	
Length and width	13 ft 6 in. by 10 ft (4.1 m by 6.1 m)
Slope	45°
Spacing of bars	6 in. (152 mm)
Estimated % ore through grizzly	65%
Open side setting	$5\frac{1}{2}$ in. (140 mm)
Eccentric throw	$1\frac{1}{2}$ in. (38 mm)
Crusher drive	Belt drive
Capacity of crushed ore pocket	None
Crushed ore feeder	6 ft by 30 ft apron, variable speed (1.83 m by 9.15 m)
Takeaway belt	54 in. (1.37 m)
Operating data	
Shifts operated per week	21 (down 1 shift every two weeks)
Effective operating hours	8
Nominal capacity	1145 stph
Maximum capacity (peak)	1200 stph
Effective availability	80%
Backing for liners	Zinc
Mantle life	6 months
Concave life	2 years
Time to change mantle	$4\frac{1}{2}$ hours
Time to change concaves	16 hours (12 men)
Liner materials in chutes	Manganese steel
Capacity of overhead crane	30 tons
Capacity of fine ore bin after crusher	12,000 st
Direct operating cost per ton (Inlcudes labor, power, supplies, maintenance)	$0.058/ton (January 1977)

54 INCH GYRATORY CRUSHERS

MINE	Anaconda/Twin Buttes	Bougainville Copper	Climax Molybdenum	Inspiration	Kennecott/Ray
Location	Sahuarita, Arizona	Papua New Guinea	Henderson, Colorado	Ray, Arizona	Ray, Arizona
Ore type	Copper porphyry	Copper porphyry	Molybdenum, granitic	Copper porphyry	Copper porphyry
Crusher size	(2) 54 in. by 80 in.	(2) 54 in. by 74 in.	54 in. by 84 in.	54 in. by 74 in.	54 in. by 74 in.
Capacity -dump pocket	73 cu yd	Nil	Not applicable	130 tons	200 tons
Crusher feeder	Vibrating	None	None	None	None
Grizzly	284 in. by 96 in. @ 10 in.	24 ft by 36 ft - 8 in.	17 ft by 46 ft	None	None
Removing oversize	Rock grapple	Rock grapple	Rock hook	Crane and rock grapple	10 t crane and breaker
Product size	$8\frac{1}{2}$ in.	6 in.	8 to 9 in.	7 in.	8 in.
Nominal capacity	4500 stph	3500 stph	3000 stph	1500 stph	1200 stph
Capacity crushed ore pocket	50 cu yd	500 st	660 st	250 st	120 st
Crushed ore feeder	Vibratory	Two 72 in. aprons each	84 in. apron	72 in. by 20 ft apron	72 in. belt v. s. to 200 fpm
Takeaway belt	60 in. @ 667 fpm	54 in. @ 650 fpm	60 in. @ 480 fpm	42 in. @ 500 fpm	54 in. @ 450 fpm
Capacity service crane	35 t mobile	60 t/10 t	75 t/15 t	75 t mobile	60 t/10 t
Direct operating cost/ton	N. A.	N. A.	N. A.	$0.09 May 1977	N. A.
Concrete used	N. A.	9,500 cu yd	9,475 cu yd	1550 cu yd	N. A.
Reinforcing and structural steel	N. A.	2696 t	963 t	N. A.	N. A.
Overall height installed	68 ft 4 in.	102 ft 3 in.	111 ft	76 ft 6 in.	61 ft 0 in.
Installation cost	$1,100,000 (1968)	ca $17,600,000 - 1972*	N. A.	$2,146,000 (1972)	N. A.
Comments	Low construction cost, in-pit installation		Circular construction	Circular construction	Fine low-cost installation

* cost includes coarse ore tunnel and conveyor

54 INCH GYRATORY CRUSHERS

MINE	Kennecott/Bonneville	Palabora Mining Co.	Palabora Mining Co.	Similkameen
Location	Bonneville, Utah	Palabora, South Africa	Palabora, South Africa	Princeton, B. C.
Ore type	Copper porphyry	Carbonatite, foskerite, and magnetite		Copper-pyrometasomatic
Crusher size	54 in. by 74 in.	(2) 54 in. by 74 in.	54 in. by 74 in.	54 in. by 74 in.
Capacity-dump pocket	3500 t	600 t filled	330 t	300 t
Crusher feeder	Railroad cars	None	None	None
Grizzly	28 ft by 37 ft - 38° $4\frac{1}{2}$ in. by 6 in.	None	None	None
Removing oversize	Crane	Crane, rock hook	Crane, rock hook	Hydraulic picker
Product size	6 in.	8.1 in.	8.1 in.	8.5 in.
Nominal capacity	2000 tpd	1265 tph each	1265 tph	1500 tph
Capacity crushed ore pocket	500 t	150 t	441 t	165 t
Crushed ore feeder	72 in. by 19 ft apron	72 in. apron	96 in. by 20 ft belt	96 in. by 50 ft belt
Takeaway belt	60 in. @ 300 fpm	60 in.	60 in.	48 in. @ 450 fpm
Capacity service crane	75 t/15 t	75 t/15 t	80 t/20 t	82 t mobile
Direct operating cost/ton	N. A.	$0.048 4 mo. 1976	N. A.	$0.05 1975
Concrete used	N. A.	8566 cu yd	7680 cu yd	2200 cu yd
Reinforcing and structural steel	N. A.	764 t	480 st - rebar only	186 t
Overall height installed	139 ft 0 in.	74 ft 8 in.	101 ft 1 in.	59 ft 0 in.
Installation cost	N. A.	N. A.	N. A.	ca $1,400,000 - 1972*
Comment	Height due to railroad cars dumping over grizzly		Figure 8 design	Fine, low-cost installation *cost excludes crushed ore conveyor

ANAMAX

ANAMAX MINING COMPANY
TWIN BUTTES MINE
SAHUARITA, ARIZONA

Copper ores from small mines in the area south of Tucson, where the Twin Buttes mine is located, were first operated in the 1880's. Today this area is one of the great copper mining centers of the world. In 1963, Anaconda Company acquired the property from Banner Mining Company and in recent years joined with AMAX in a 50/50 partnership to work the Twin Buttes property.

From the beginning, Anaconda had relied on conveyors to remove overburden, oxide material for eventual leaching, and ore from the open pit. In the largest preproduction stripping program in mining history, Anaconda in a four-year period (1965 to 1969) removed 266 million tons of overburden of an average depth of 500 ft (152 m) lying over the ore body. The alluvium overburden was stripped by scrapers carrying 70 to 80 tons and push-loaded by two D9 tractors operating in tandem. The scrapers delivered their loads to bins which fed a 60 in. (1.52 m) conveyor having a capacity of 8500 tph. The 25% slope belt conveyor, which was 2100 ft (640 m) long, transported the material to the brow of the pit, where it was transferred to an 8300 ft (2530 m) surface belt to a 1000 ton loadout bin. At this point, the overburden was trucked to the sites of the tailings ponds, where it was used for dam construction. During 1976, the one billionth ton of overburden was removed from the Twin Buttes mine. The Twin Buttes quartz monzonite porphyry contains only scattered mineralization of which chalcopyrite is the most abundant mineral, associated with chalcocite, covellite, minor bornite, and several oxide minerals. Oxidation has progressed to variable depths of up to several hundred feet below the base of the alluvium.

Because of the depth of the sulfide and oxide ores and hard rock waste, two in-pit crushers and conveyor systems were also installed for their transport out of the pit; and in 1976 a third crusher and conveyor system was installed lower in the pit for waste rock only. The two initial crushing and conveying systems were designed to handle any of three functions: crushing waste to be disposed of by a rail-mounted stacker-conveyor, crushing oxide ore for conveyor and then truck transport to stockpiles for future leaching, and primary crushing of sulfide ore for conveyor transport to the two 30 in. by 70 in. (0.76 by 1.78 m) secondary crushers and the two 7 ft (2.13 m) standards and four 7 ft shortheads in the third and fourth stage crushing plant. A third standard crusher has been installed recently.

Mining is done by 15 yd (11.5 cu m) shovels which load 100- and 150-ton trucks for transport of the sulfide ore, oxide ore, or waste to the two 54 in. by 70 in. Nordberg gyratory crushers located on benches in the pit wall. At each crushing station, the trucks dump one at a time onto a 4 ft (1.22 m) stationary grizzly mounted over a 73 cu yd (56 cu m) pocket equipped with an Eliptex 2-E-13 vibrating grizzly feeder. Boulders larger than 4 ft (1.22 m) are removed by a hydraulic grapple or a front-end loader equipped with heavy tines on the bucket lip for secondary blasting. The last third of the feeder (95 in. or 2.4 m) is composed of grizzly bars set at 10 in. (254 mm) which pass about 65% of the ore. Grizzly oversize falls only a few inches to the crusher feed opening and across the crusher spider, which is parallel to the axis of the grizzly.

Crushed ore at $8\frac{1}{2}$ in. (216 mm) is discharged into a 50 cu yd (38.2 cu m) pocket where it joins the grizzly undersize and is fed by an Eliptex 2-E-13 feeder onto a short, 60 in. (1.52 m) belt which in turn feeds the main 60 in. conveyor for transport to the rim of the pit. Each of the two crushing systems from the truck dump to the pit rim are mirror images of each other. Including the grizzly undersize, each crushing installation has a 4500 stpd nominal capacity and a maximum capacity of 6000 stph.

The crushers are installed in the pit wall and concrete is used only up to the crusher floor. Above that point, structural steel is used. This design was utilized because as the pit was deepened the crushers could be moved to a lower depth in the pit. Both crushers are installed entirely in the open, and no permanently mounted crane was installed. During relining of the crushers, or to pull the main shaft, an 85-ton mobile crane is brought into service. Relining is done with epoxy instead of zinc, which can be dangerous where there is a possibility of it contacting moisture. Another feature is the use of a belt drive and high-speed motor instead of the more commonly used direct drive. Operating and maintenance of the crushers in the open presents no particular problems.

The 4500 and 6000 stph nominal and maximum capacities of these crushers are due to the way the ore breaks in the pit and, even more importantly, to the 10 in. (254 mm) setting of the vibrating grizzly; and the latter is permitted by the four-stage crushing flowsheet, which is unusual for a copper ore. It is interesting to speculate how the crushers would operate without the primary mechanical vibrating grizzly feeder, a high-maintenance piece of equipment which cannot be bypassed. Deletion of the feeder would, of course, have introduced substantial structural problems in supporting the weight of a loaded truck on the pit wall. In any event, Anaconda's in-pit crushing scheme with conveyor transportation to the surface has been the forerunner of a new technology for deep open-pit mining. As stated elsewhere, Duval has recently installed a twin in-pit crushing plant at their Sierrita operation only a few miles away.

The capital cost of each of the ANAMAX crushers in 1966 was $1,100,000.

ANAMAX MINING COMPANY

Sahuarita, Arizona

Ore type	Copper-porphyry skarn, 148#/cu ft (2371 kg/m^3)
Concentrator capacity	40,000 st/day
Shovel size	15 yd (11.5 m^3)
Truck sizes	100 and 150 st
Primary crusher	Two 54 in. by 80 in. Nordberg gyratories
Connected and consumed horsepower	500 - 375 at 700 rpm
Capacity of dump pocket	73 cu yd (55.8 m^3)
Crusher feeder	Hewitt Robins Eliptex 2 - E-13 vibrating grizzly
Length and width of grizzly section	284 in. by 96 in. (7.24 m by 2.44 m)
Slope	6°
Spacing of bars	Approximately 10 in. (254 mm)
Estimated % ore through grizzly	65%
Method of breaking or removing oversize	Rock grapple
Open side setting (product size)	8½ in. (216 mm)
Crusher drive	Vee belt
Capacity of crushed ore pocket	Approximately 50 cu yd (38.2 m^3)
Crushed ore feeder	Eliptex E-13 vibratory; material movement approximately 60 fpm
Takeaway belt	60 in. at 667 fpm (1.52 m at 203 mpm)
Operating data:	
Shifts per week	15
Effective operating hours	6.75 per shift
Nominal capacity	4,500 st/hour
Maximum capacity	6,000 st/hour
Effective availability	76 to 85% of scheduled time
Backing for liners	Epoxy
Mantle life	6 months
Concave life	6 months
Time to change mantle	16 hours
Time to change concaves	60 hours
Liner materials in dump pocket and crushed ore pocket	T-1-360 Brinell, 1 in. thick (25.4 mm)
Capacity of overhead service crane	35-ton mobile crane
Coarse ore stockpile (after crusher)	67,000 t live, after secondary crushing
Direct operating costs	Not available
Capital cost	$1,100,000 each (1968)

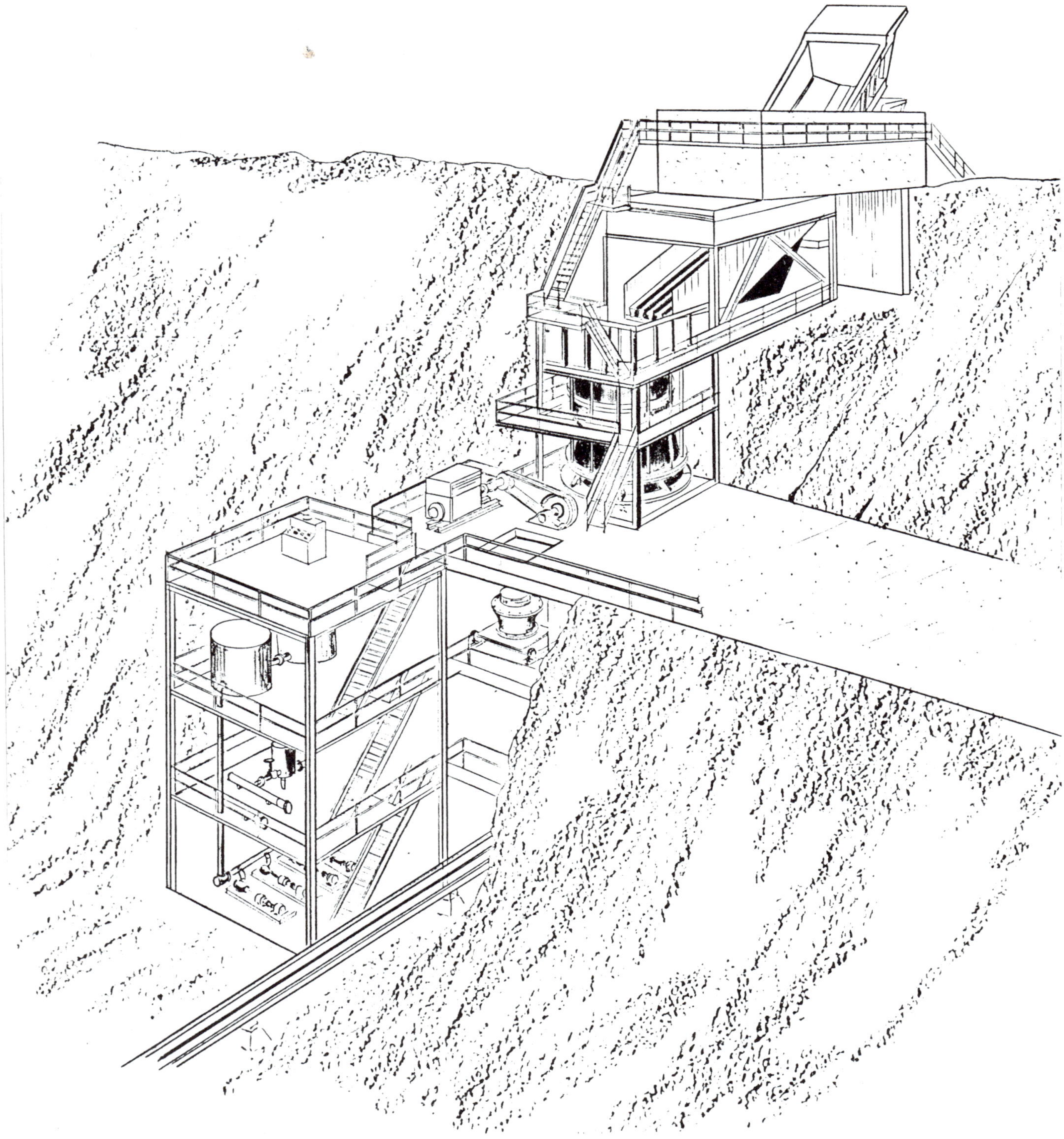

ANAMAX MINING CO.
TWIN BUTTES MINE
SAHUARITA, PIMA COUNTY ARIZONA

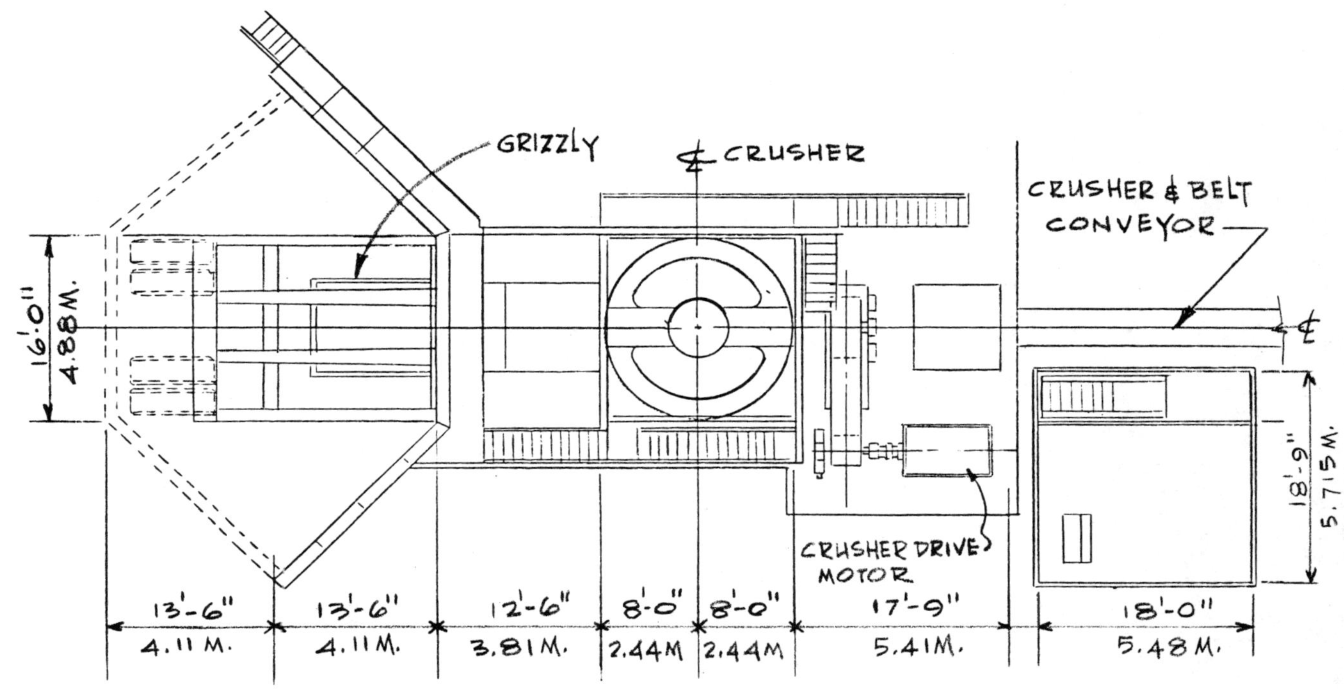

PLAN

ANAMAX MINING CO.
TWIN BUTTES MINE
SAHUARITA, PIMA COUNTY ARIZONA

BOUGAINVILLE COPPER COMPANY
PANGUNA, BOUGAINVILLE
PAPUA NEW GUINEA

The Bougainville ore body was discovered in 1964 by a partnership of Conzinc Rio Tinto of Australia and New Broken Hill Consolidated Limited. The ore body is a porphyry copper in a fine-grained quartz diorite, which intrudes andesite volcanics and is located at an elevation of about 2,000 ft in the Crown Prince Range of Bougainville in the Solomon Islands. Mineralization is predominantly chalcopyrite with some bornite and some gold and silver present. Rainfall in the area amounts to 200 in. (5.08 m) per year. An active volcano is situated only 24 miles (38.4 km) from the mine, and all structures were designed to resist seismic shocks. An interesting and unique feature of the Bougainville operation was that the overburden was stripped with hydraulic monitors.

Bougainville Copper, at 90,000 tpd, is the largest copper concentrator to be designed and constructed at one time; and besides the twin 54 in. by 74 in. Allis-Chalmers gyratories, it features seven 7 ft (2.13 m) standard and fourteen 7 ft shorthead crushers, in addition to nine 18 ft diameter by 21 ft long (5.5 m by 6.4 m), 4250 hp, single-stage ball mills.

During the design stage of the Bougainville primary crushing plant, several alternates were studied because the ore was expected to break with the production of large amounts of small-sized particles but with a minimum of fines. These alternates included two 54 in. by 74 in. gyratories with grizzlies, two 60 in. by 89 in. gyratories, and one 72 in. by 96 in. gyratory, the latter two installations without grizzlies. The two 54 in. units with grizzlies were finally selected as giving good reliability and the ability to handle the large amounts of small material in the feed. As originally installed, grizzly bars were set at 8 in. (203 mm); but, in an effort to perform more interparticle crushing in the primary and less rock-on-steel crushing in the secondaries and tertiaries, a test is underway (1976) to determine the effect of using 6 in. (152 mm) grizzly spacings (see write-up on Anaconda's crusher at Butte).

The Bougainville ore is mined with 17 cu yd (13 cu m) shovels and hauled in 105, 170, and 210 st trucks to the crushers, which are equipped for two-side dumping. Up to 40% of the ore passes the grizzlies, which are equipped with splitters; and additional horizontal splitter bars are located directly above the spider caps. Because of the grizzlies, the crushers are supported by unique U-shaped enclosures, which are in turn supported by the end and center walls of the installation.

The crushed ore pockets are necessarily wide to accomodate the grizzly undersize and hold 500 st each. Two 6 ft wide (1.83 m) variable-speed apron feeders under each crusher transfer the crushed ore to two parallel 54 in. (1.37 m) takeaway belts which serve both crushers. With one apron feeder or belt conveyor down for maintenance, a crusher may still operate by dumping ore only into the opposite side. Radioactive level indicators on each side of the crushed ore pocket control the corresponding truck dump lights.

Maintenance access to the crushers is through hatches between them, as is usual with two-crusher installations. This area was equipped with stairways and space for an elevator to be added later. Eccentric removal is eased by a unique double drawbridge arrangement. Doors on either side of the crushed ore pockets are hinged at their bottoms. Eccentric cart rails in the outer door faces join when the doors are lowered into pockets.

On the surface, the electrical room is located at one end of the installation and the repair area with mainshaft stands is located at the other. Handling of the mainshafts and maintenance materials is done by a bridge crane equipped with 60- and 10-ton hoists. Both the concaves and mantles are secured with epoxy resin instead of zinc. Dust collection piping is located between the crushers, and a wet collector is situated partially above and below the floor at elevation zero. Since the ore is already wet from rainfall, the collected and slurried dust is pumped to the concentrator. No water sprays are used in the dump pocket.

Although the double grizzlies contributed substantially to the overall height of this fine crushing plant, concrete usage was only 9,500 yd (7264 cu m). Total steelwork, including the superstructure, ductwork, liners, and rebar, amounted to 3,500 tons.

BOUGAINVILLE COPPER COMPANY

Papua New Guinea

Ore type	Copper - Porphyry, 93#/cu ft (1538 kg/m^3)
Concentrator capacity	86,000 st/day
Shovel size	17 yd (13 m^3)
Truck sizes	105, 170, and 210 st
Primary crusher	Two 54" x 74" Allis Chalmers Gyratories
Connected and consumed horsepower	500 - 50 ave.
Capacity of dump pocket	Nil
Grizzlies	Double
Length and width	24 ft and 36 ft (7.3 and 11 m)
Slope	37 degrees
Spacing of bars	8 in. (6 in. under test) 203-152 mm)
Estimated % ore through grizzly	20% - 40%
Method of breaking or removing oversize	Grapple hook
Open side setting (product size)	6 in. (152 mm)
Eccentric throw	1 5/8 in. (41.5 mm)
Crusher drive	Direct
Capacity of crushed ore pocket	500 st
Crushed ore feeder	Two 6' wide apron feeders per crusher (1.83 m)
Speed of feeder	2.6 to 27.0 fpm (0.8 - 8.3 mpm)
Takeaway belt	Two 54" wide; 615 fpm. One belt sped up to 780 fpm during 1975; other belt to be sped up in 1976 (1.37 m-187 mpm)
Operating data:	
Shifts operated per week	21
Effective operating hours	Utilization 68% - 5.4 hours per shift
Nominal capacity	3,500 st/hour - total
Maximum capacity (peak)	6,500 st/hour - total
Effective availability	85% of scheduled time
Backing for liners	Epoxy
Mantle life	12 months
Concave life	12 months
Time to change mantle and concaves	48 hours
Capacity of overhead service crane	60 ton - main; 10 ton auxiliary
Coarse ore stockpile (after crusher)	50,000 tons live; 150,000 tons total
Direct operating cost per ton	Not available
Concrete	9,500 cu yd (7264 m^3)
Reinforcing bar	940 tons
Embedded steel	155 tons
Structural steel	1,100 tons
Miscellaneous steel:	
Plate and structural	501 tons
Liners	322 tons
Wear bars	258 tons
Ductwork	223 tons
Overall height of installation	102 ft $2\frac{1}{2}$ in. (31/15 m)
Constructed cost (includes tunnel and conveyor)	ca $17,600,000 (June 1972)

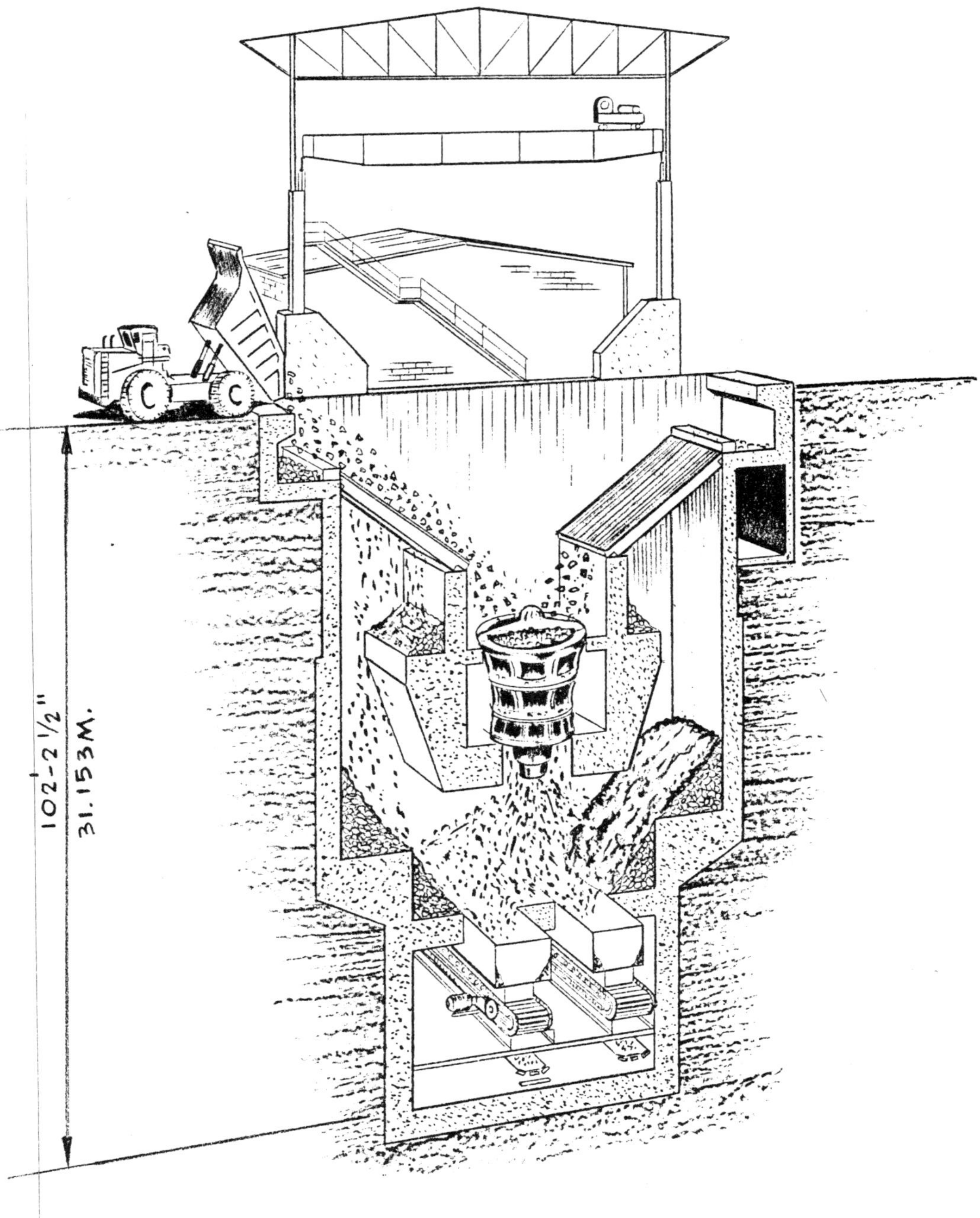

BOUGAINVILLE COPPER CO.
SOLOMON ISLANDS

CLIMAX MOLYBDENUM COMPANY
HENDERSON, COLORADO

The Climax Molybdenum Company's newest operation is the Henderson project, which is located on the east side of the Continental Divide. Ore reserves are estimated at more than 300 million tons at a grade of 0.49% MoS_2. For mining the ore a modified panel caving system is used. The ore is transported to the Henderson mill on the west side of the Continental Divide on a 42 in. (1.07 m) gauge railroad through 9.6 miles (15.4 km) of tunnel and 4.8 miles (7.7 km) of surface track, a total distance of 14.4 miles (23.2 km). Each train consists of 30 ore cars with capacity of 22 st (20 mt) of ore or 660 st (599 mt) per train. The unmanned trains operate completely automatically.

At the 54 in. by 84 in. primary gyratory crusher, the bottom dump ore cars pass over a dump scroll to discharge at a maximum rate of 100 st (91 mt) per minute. Dumped ore falls onto a rock box before passing over a 17 ft long by 46 ft wide (5.2 m by 14 m) grizzly set at a 40° angle with bar spacings of $7\frac{1}{2}$ and $8\frac{1}{2}$ in. (191 mm and 216 mm) openings. An estimated 30 to 50% of the ore (which at times is wet and sticky) passes through the grizzly. The operators report that a shallower angle on the grizzly, perhaps 35° to 37°, would have been sufficient and would have prevented impact on the back of the crusher pocket by large boulders. This pocket as well as the crushed ore pocket under the crusher is lined with AR plate. The accompanying drawings show a "future splitter" and the operators are planning the installation of this unit to prevent impact and subsequent damage to the crusher spider.

Crushed ore at an 8 to 9 in. (203 mm to 229 mm) size joins the grizzly undersize in the pocket beneath the crusher. This pocket is sized at 660 live st (599 mt) so that it will accept an entire trainload of ore. Crushed ore is removed from the pocket by a four-speed, 84 in. (2.1 m) apron feeder operating at 9.6 to 29.3 fpm (2.9 to 8.9 mpm). A 60 in. (1.5 m) takeaway conveyor operating at 480 fpm (146 mpm) delivers the crushed ore to a 58,000 st live and 193,000 st total (52,600 and 175,100 mt) open stockpile under an A-frame supported tripper gallery. Impact idlers on the takeaway conveyor under the apron feeder discharge are mounted on wood blocks for a cushioning effect and ripped belt detecting bars are located behind each impact idler. The conveyor tunnel is of corrugated and galvanized steel construction rather than concrete.

The concrete crusher enclosure is circular in design from its lowest level at elevation minus 110 ft 4 in. (minus 33.6 m) to elevation minus 37 ft 6 in. (minus 11.4 m) because it was located in solid rock and the circular excavation minimized the rock removed. Both the switchgear room and the crusher lubrication equipment room are located in this circular section, which is 88 ft (26.8 m) in inside diameter. Crusher eccentric removal is accomplished by a cart suspended on widely spaced rails permanently mounted near the top of the crushed ore pocket, so the crushed ore will fall between the rails during operation. Monorail trolley beams are conveniently located to aid servicing the crusher drive system and the apron feeder. Spile rod guides have been provided above the feeder during repair. Access to all levels is by a large service elevator.

Above elevation minus 37 ft 6 in. (minus 11.4 m), the crusher enclosure changes in shape from circular to square where an additional room has been added on the outside of the square to accommodate the dust collection system and air heating units. Above the dumping level the entire enclosure is covered by a building equipped with a 75/15 ton crane which is used for maintenance and for removing an occasional oversize boulder from the dump pocket. A generously sized area is also provided for mantle relining. Mantles and concaves are backed with zinc. Roll-up doors permit this building to be heated when the ore trains are not dumping. Above the crusher pocket is a control room with an excellent view of the dumping and crushing operations. Also in the control room is the train control system which shows the location of all the trains being operated. Capacity of this crushing plant is 3,000 stph nominal and 6,000 stph maximum (2,722 and 5,433 mtph).

Although the 88 ft (26.8 m) diameter of the enclosure produces an extremely large volume, a substantial portion of which is not used, the quantity of concrete (9,475 cu yd or 7.245 m^3) and the amount of reinforcing, structural, and miscellaneous steel were relatively low since the circular design permits such economies.

CLIMAX MOLYBDENUM COMPANY

Henderson, Colorado

Ore type	Molybdenum - granitic, 105#/cu ft (1682 kg/m^3)
Concentrator capacity	30,000 stpd (27,200 mtpd)
Shovel size	Underground mining uses 5 cu yd (3.8 m^3)LHD's
Rail car size	22 st (20 mt)
Primary crusher	54 in. by 84 in. Traylor gyratory
Connected and consumed horsepower	500 and 200-400 (377 and 150-300 kw)
Capacity of dump pocket	Not applicable
Grizzly	Single
Length and width	17 ft and 46 ft (5.2 and 14 m)
Slope	40°
Spacing of bars	$7\frac{1}{2}$ and $8\frac{1}{2}$ in. (191 and 216 mm)
Estimated % of ore through grizzly	30-50%
Method of breaking or removing oversize	Rock hook on crane
Open side setting (product size)	8 - 9 in. (203 - 229 mm)
Eccentric throw	$1\frac{1}{2}$ in. (38 mm)
Crusher drive	Direct
Capacity of crushed ore pocket	660 st (599 mt)
Crushed ore feeder	84 in. wide (2.1 m) apron
Speed of feeder	9.6, 14.5, 19.4, 29.3 fpm (2.9, 4.4, 5.9, 8.9 mpm)
Takeaway belt	60 in. - 480 fpm (1.52 m - 146 mpm)
Operating data·	
Shifts operated per week	20
Effective operating hours	8 per shift
Nominal capacity	3,000 stph (2,722 mtph)
Maximum capacity (peak)	6,000 stph (5,443 mtph)
Effective availability	+95%
Backing for liners	Zinc
Mantle life	Unknown
Concave life	Unknown
Time to change mantle	12 hours, estimated
Time to change concaves	60 hours, estimated
Capacity of overhead service crane	75 and 15 ton
Coarse ore pile (after crusher	58,000 st live-193,000 st total (52,600 and 175,100 mt)
Direct operating cost	Not available
Concrete	9,475 cu yd (7,245 m^3)
Reinforcing steel	612 st (555 mt)
Structural steel	351 st (318 mt)
Miscellaneous steel (embedded metal, wear bars, etc.)	340 st (308 mt)
Overall height of installation	111 ft (33.83 m)
Date of completion	July 1976

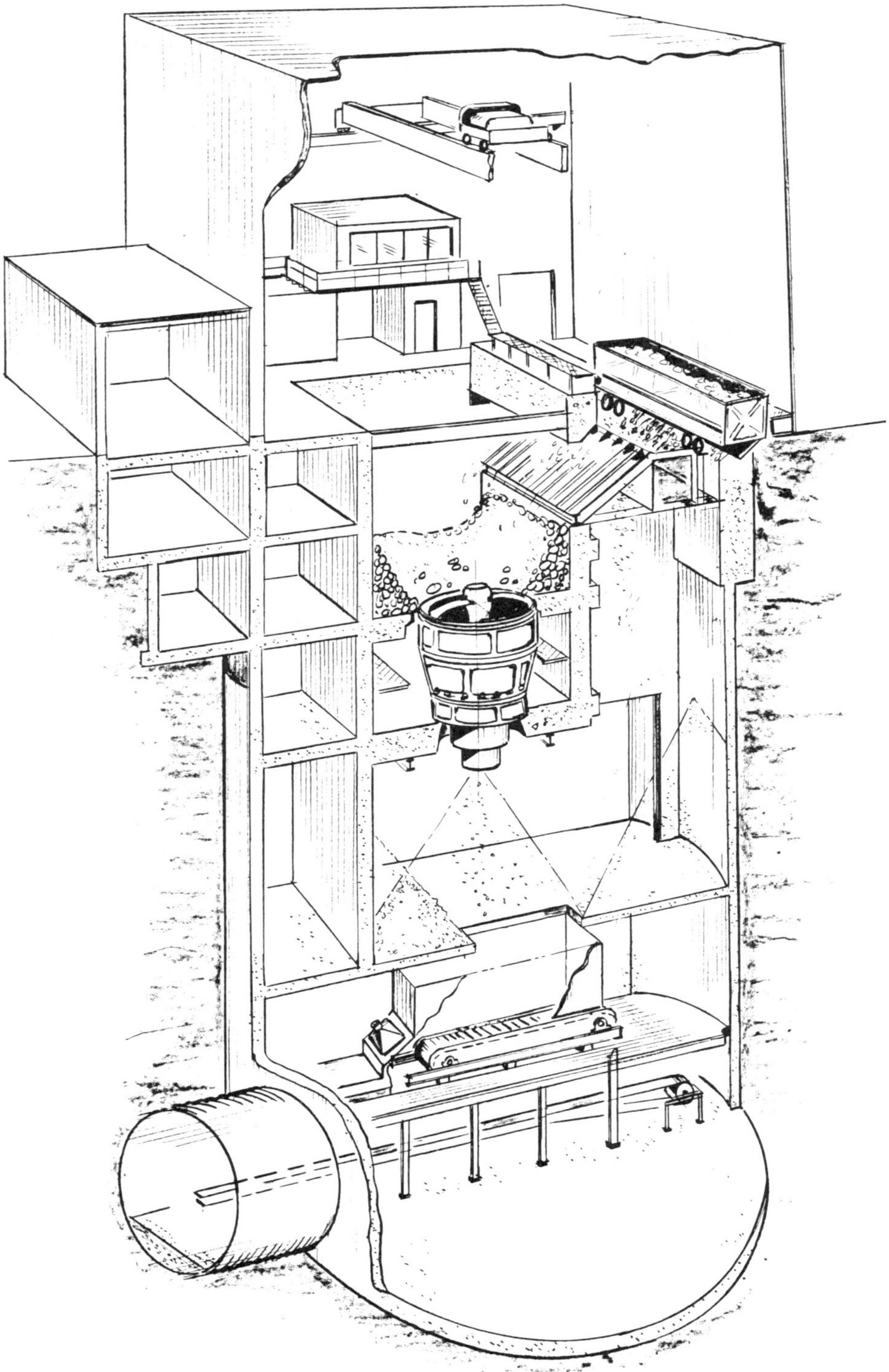

CLIMAX MOLXBDENUM CO.
HENDERSON, COLORADO

INSPIRATION CONSOLIDATED COPPER COMPANY
INSPIRATION, ARIZONA

The Inspiration ore body was discovered in 1904, but a 50-ton mill which was installed to treat the disseminated copper ore promptly failed. The original Inspiration Company then gave way to another, backed by W. B. Thompson, which found 40 million tons of ore. Control then passed to Anaconda and Dr. Louis Ricketts took charge as consulting engineer. He abandoned the mill which had been partially built (to the horror of the stockholders, according to Ira Joralemon) and spent a year and a million dollars experimenting with flotation. The first large copper flotation plant was then built and recovered 50% more copper than had previous gravity mills. It, of course, was highly successful and became the forerunner of modern flotation plants.

Inspiration is distinctive as the first fully integrated copper operation in the United States. Three open pits supply sulfide ore to the concentrator and mixed oxide and sulfide ore to a vat leaching and electrowinning operation with leached ore also being sent to the concentrator. In addition, dump leaching supplies pregnant solution for scrap iron precipitation. Inspiration has recently constructed an electric furnace smelter and also operates a continuous casting and rod plant.

The new primary 54 in. by 74 in. crusher was completed in May of 1972 and replaced an older unit which was in the way of the open pit development. It is of an unusual circular silo design, which permits dumping from two positions at 90° from each other. One dumping position is in line with the spider, while the other one is at right angles to it. A box girder splitter bar hung from the dump pocket slopes downward from the second dump point, across the spider cap, and rests on the pocket bottom. This splitter was not installed to protect the spider but was to aid in distributing wet ore being dumped from that position to both sides of the spider. Any large amount of wet ore delivered to one side of the spider would pack in the crusher cavity. The round front wall of the 18 ft (5.5 m) deep pocket is unlined except for light steel protection at the top, while the straight back wall is lined with vertically positioned railroad rails. Water for dust suppression is sprayed into the pocket through a cluster of 3/8 in. (9.5 mm) nozzles at the back of the pocket. About 5 gpm (19 lpm) of water are mixed with 90 psi (620 kPa) air upstream of the nozzles to produce a fog which is effective in dust control.

The 54 in. Allis-Chalmers gyratory is placed mostly within one quadrant of the circular enclosure to permit placement of the apron feeder on the lowest level in the proper position to transfer crushed ore to the takeaway conveyor. Because of the 40 ft (12.2 m) diameter of the circular enclosure, the crusher could not be supported by a conventional diaphragm but instead is supported by two vertical walls spanning the enclosure. These walls also form the tall but narrow crushed ore pocket. The bottom half of the pocket is longer than the top half, which permits the crusher eccentric to be removed through doors in the latter. The eccentric and cart must then be turned 90° and moved to the hoistway for removal. Because wear on the upper half of the crushed ore pocket is extreme, the walls are covered with 2% chrome manganese steel. Two radioactive level indicators are used in the crushed ore pocket; one is about two feet (0.6 m) above the apron feeder and the other is at a level which permits the ore to come within 18 in. (457 mm) of the hydroset mechanism.

Inspiration has dispensed with the conventional bridge crane over the crusher, as did Similkameen and one or two of the cement companies, and uses a 75-ton and a 35-ton mobile unit to lift the main shaft. Mantle changeout requires three to five 12-hour shifts, depending on the number of mechanical replacements needed; and concave relining requires five to eight 12-hour shifts.

Oversize rocks were originally broken with a hydraulic hammer mounted on an articulated arm, but because of the difficulty encountered in holding the hammer in place, the arm is now being equipped with a hook for turning the rocks. The rare oversize piece will be removed by a mobile crane. Again, as at Similkameen, no elevator was provided, although the total height of the crusher enclosure is 76 ft (23.2 m), 17 ft (5.18 m) more than at Similkameen. Electrical equipment is installed on the floor around the crusher instead of in a separate electrical room.

Dust piping is located in the area under part of the dump pocket. Dust is drawn downward to the base of the silo to a pipe which runs under the horizontal conveyor in the bottom of the corrugated metal pipe enclosing the conveyor. At the daylighted end of the pipe is a picking station, where wood and iron can be removed from the crushed ore and dropped into a dump truck for disposal. The ore is also transferred at this point to a takeaway conveyor which delivers it to the coarse ore bins. A dry cyclone and a wet multiclone are located over the takeaway conveyor and deposit the collected dust on the crushed ore on the conveyor.

The circular design of the crusher enclosure required only 1550 cu yd (1185 cu m) of concrete; this figure includes wing walls at the takeaway conveyor tunnel exit. Capital cost of the crusher (May 1972) was $2,150,000. Operating costs, including labor, power, supplies, and maintenance, were $0.09 per ton in December 1976. This cost included the operating of the takeaway conveyor, its tripper over the crushed ore bins, the bins themselves, and the 24 vibrating feeders underneath them.

INSPIRATION CONSOLIDATED COPPER COMPANY

Inspiration, Arizona

Ore type	Copper porphyry, 111 #/cu ft (1778 kg/m^3)
Concentrator capacity	20,000 stpd
Shovel size	6, 8, and 10 yd (4.6, 6.1, and 7.6 m^3)
Truck size	85 st
Primary crusher	54 in. by 74 in. Allis-Chalmers
Connected and consumed horsepower	450 - ± 250
Capacity of dump pocket	1½ truck loads used
Grizzly	Not used
Method of breaking or removing oversize	Rock manipulator - mobile crane
Open side setting (product size)	7 in. (178 mm)
Eccentric throw	1 3/8 in. (40 mm)
Crusher drive	Direct through clutch
Capacity of crushed ore pocket	250 st
Crushed ore feeder	72 in by 20 ft apron (1.83 by 6.1 m)
Speed of feeder	33 fpm fixed (10 mpm)
Picking belt	72 in. @ 209 fpm (1.83 @ 63.7 mpm)
Takeaway belt	42 in. @ 500 fpm (1.07 m @ 152 mpm)
Operating data:	
Shifts operated per week	10
Effective operating hours	7¼ per shift
Nominal capacity	1500 stph
Maximum capacity (peak)	2000 stph
Effective availability	95%
Backing for liners	Zinc
Mantle life	12,000,000 st
Concave life	18,000,000 st
Time to change mantle and concaves	3 to 5 12-hour shifts and 5 to 8 12-hour shifts
Capacity of overhead service crane	75 ton mobile
Coarse ore bins (after crusher)	7500 st
Liner materials	Dump pocket: dead bed; ore pocket: 2% CrMn and dead bed
Direct operating cost per ton	$0.09 (includes coarse ore bins) (May 1977)
Concrete	1550 cu yds (1185 m^3)
Overall height of installation	76 ft 6 in. (23.3 m)
Capital cost	$2,146,000 (1972)

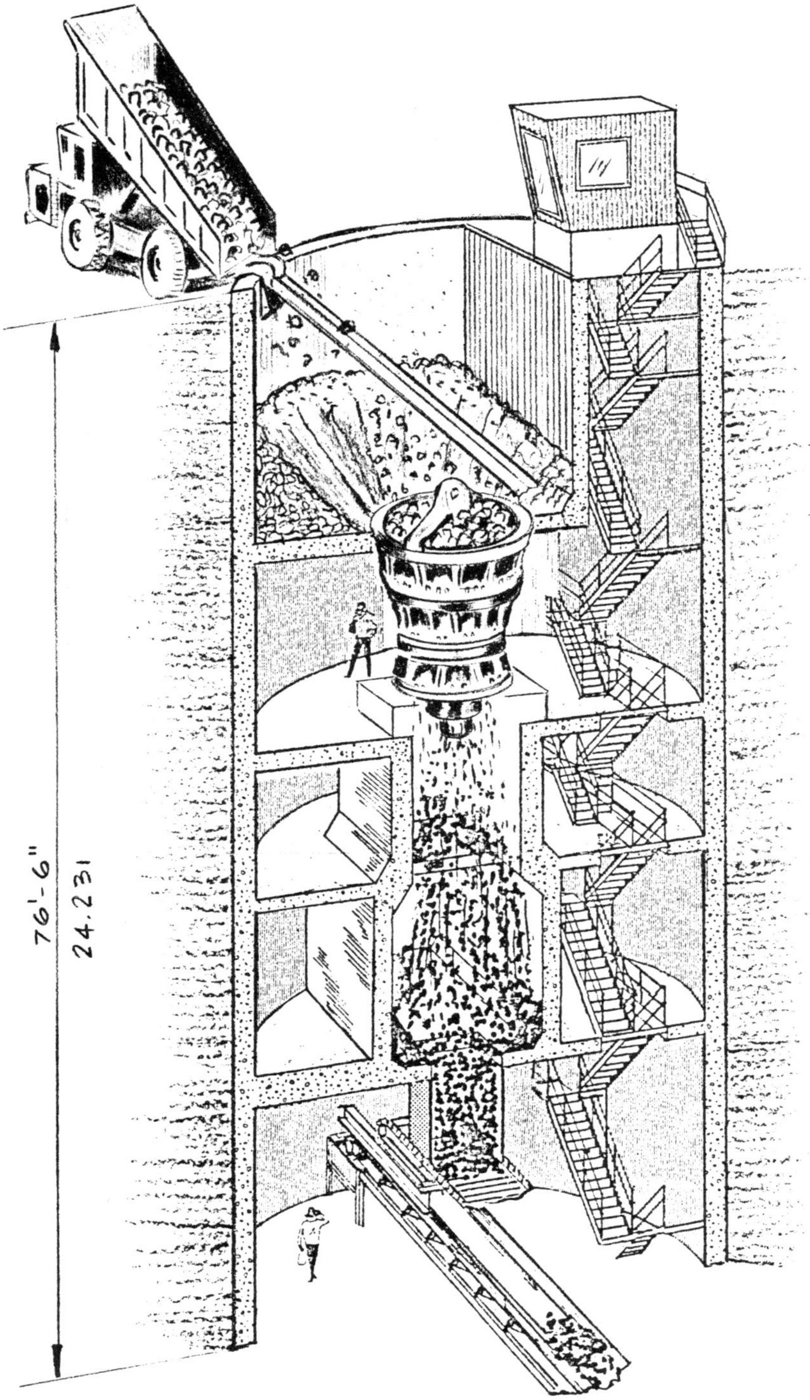

INSPIRATION CONSOLIDATED COPPER CO.
INSPIRATION, ARIZONA

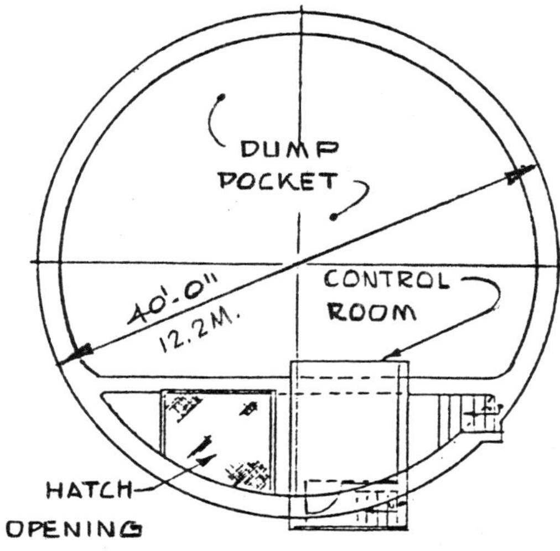

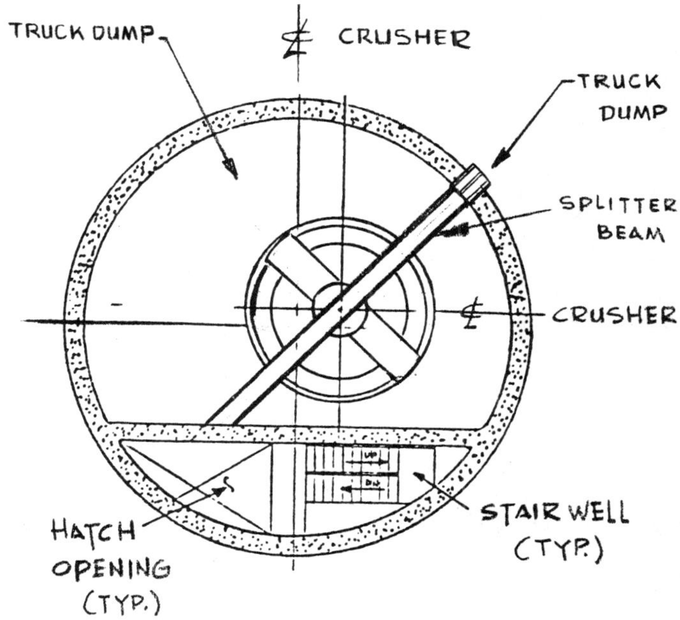

PLAN AT TOP OF SILO
ELEV. 0'-0"

PLAN AT DUMP FLOOR
ELEV. - 18'-0" = 5.486M

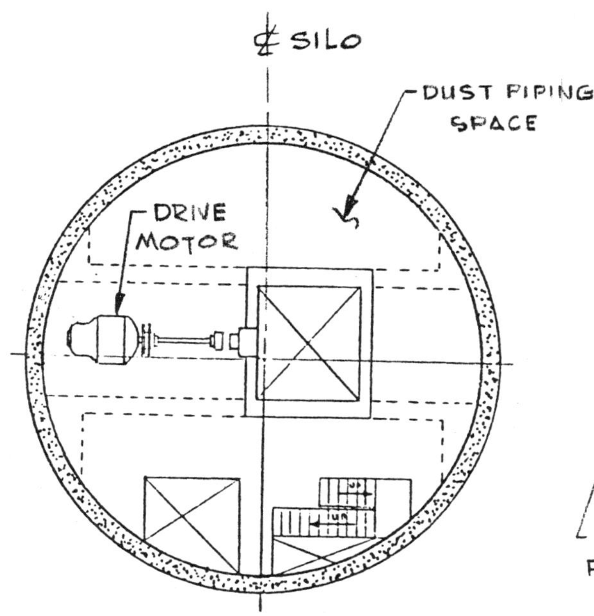

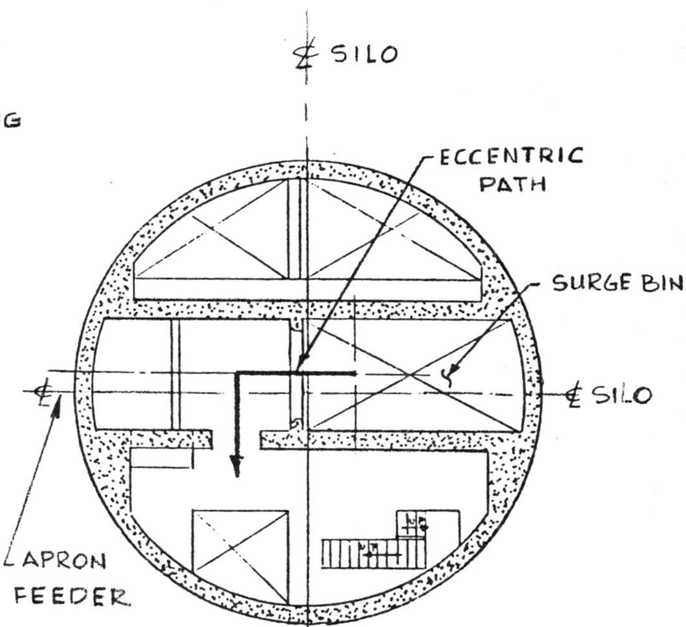

PLAN AT DRIVE MOTOR FLOOR
ELEV. 36'-6" = 11.125 M

PLAN AT HYDROSET FLOOR
ELEV. - 46'-0" = 14.021M

INSPIRATION CONSOLIDATED COPPER CO.
INSPIRATION, ARIZONA

KENNECOTT BONNEVILLE

KENNECOTT COPPER CORPORATION
BONNEVILLE PLANT
SALT LAKE CITY, UTAH

The great Bingham Canyon mine and mill complex of Magna, Arthur, and Bonneville had their beginning in a 300 ton per day operation which was commenced by Daniel C. Jackling and his associates in 1904 as the Utah Copper Company. This small plant was quickly expanded to 500 tons per day and eventually Kennecott, the Utah Copper successor, was processing about 45,000 tons per day of ore in each of its mills, Magna and Arthur. Finally, in 1966, the Bonneville crushing and grinding plant was added to produce a slurry of ground ore for flotation concentration at the expanded Magna and Arthur mills.

Bonneville is one of the few railroad dump primary gyratory crushing plants constructed in recent years; another which is described in this Monograph is Cuajone in Peru, started in 1976. Two others are the Climax Henderson unit and the Phelps Dodge Tyrone crusher, which has been fed by trucks since it commenced operation. The Bonneville, Henderson, and Cuajone crushers are all located considerable distances from their mines, hence the use of rail haulage. Shovels in the Bingham pit are 8 to 15 yd capacity; and ore is sent to the Bonneville crusher in 100 ton rail cars, where they are dumped, two at a time, by a hydraulically driven rotary dumper. Ahead of the 54 in. by 74 in. Allis-Chalmers gyratory are two grizzlies, each 27 ft 9 in. (8.46 m) long with bar spacings of $4\frac{1}{2}$ in. (114 mm) at the feed end and 6 in. (152 mm) at the discharge end. Grizzlies are sloped at 38°. Oversize rocks are removed from the crusher pocket by an overhead crane, and the crusher produces a nominal 6 in. (152 mm) product. Rail cars are dumped in the direction away from the crusher, and water sprays at the dumper level and in the rock boxes are used for dust suppression.

There are two 72 in. by 19 ft (1.83 m by 5.8 m) apron feeders under the 500 st crushed ore pocket and two more under the grizzly undersize pocket. Feeder speeds are variable up to 72 fpm. Grizzly undersize is dropped to the feeder pocket by three shelves forming a rock ladder attached to the pocket walls. Each pair of apron feeders, one for crushed ore and one for grizzly undersize, discharge to a 72 in. (1.83 m) picking belt for tramp removal; this belt feeds one of two 54 in. (1.37 m) feeder conveyors which in turn discharge to a 60 in. (1.52 m) takeaway belt operating at 175-300 fpm (53.3 - 97.4 mpm). Both picking belts are equipped with metal detectors and tramp iron magnets, and a separate 30 in. (760 mm) conveyor is used to remove tramp material.

At the surface of the crusher, the car dumper and maintenance areas are enclosed by a steel building; and doors at the track entrance and exit can be rolled down to keep the building warm for maintenance during cold weather. Crusher and dumper maintenance are facilitated by a travelling crane having 75- and 15-ton hoists.

Railroad dump crushers are necessarily deep, and Bonneville is no exception at 139 ft (42.37 m). To dump two cars at once, additional height is required to funnel the ore to the crusher opening. Grizzlies, of course, create the necessity of having more feeders or alternatively increasing the height still more to combine the grizzly undersize with the crushed ore.

Feed to the Bonneville primary crusher is 32,000 stpd, and the crushed ore stockpile contains 12,000 tons live and 60,000 tons total.

KENNECOTT COPPER CORPORATION

Bonneville Plant, Salt Lake City, Utah

Ore type	Copper-Porphyry, 110-115#/ cu ft (1610 - 1842 kg/m^3)
Production per day	32,000 st
Shovel sizes	8 to 15 yd (6.1 to 11.5 m^3)
Railroad car size	100 tons
Primary crusher	54" x 74" Allis Chalmers Gyratory
Connected and consumed horsepower	500 - consumed not available
Capacity of dump pocket	Approximately 3,500 st live
Grizzly	Double
Length and width, each	27 ft 9 in. long by 37 ft wide (8.46 by 11.3 m)
Slope	38°
Spacing of bars	4 1/2 " top, 6" bottom (114 and 152 mm)
Estimated % ore through grizzly	Not available
Method of breaking or removing oversize	Removal by crane
Open side setting (product size)	6 in. (152 mm)
Eccentric throw	1 3/8 in. (40 mm)
Crusher drive	Direct
Capacity of crushed ore pocket	Approximately 500 st
Crushed ore feeder	72" x 19' apron at 0 to 72 fpm (1.83 x 5.8 m @ 21.9 mpm)
Takeaway belt	60" at 175-300 fpm (1.52 m @ 53.3 - 91.4 mpm)
Operating data:	
Shifts operated per week	14
Effective operating hours	Not available
Nominal capacity	2,000 st/hour
Maximum capacity (peak)	2,500 st/hour
Effective availability	Not available
Backing for liners	Zinc
Mantle life	12 months
Concave life	18 months
Time to change mantle	14 hours
Time to change concaves	24 hours
Liner materials in dump pocket and crushed ore pocket	Manganese steel
Capacity of overhead service crane	75 tons, 15 tons auxiliary
Coarse ore stockpile, after crusher	12,000 live; 60,000 total
Direct operating costs per ton	Not available
Overall height of installation	139' 0" (42.37 m)

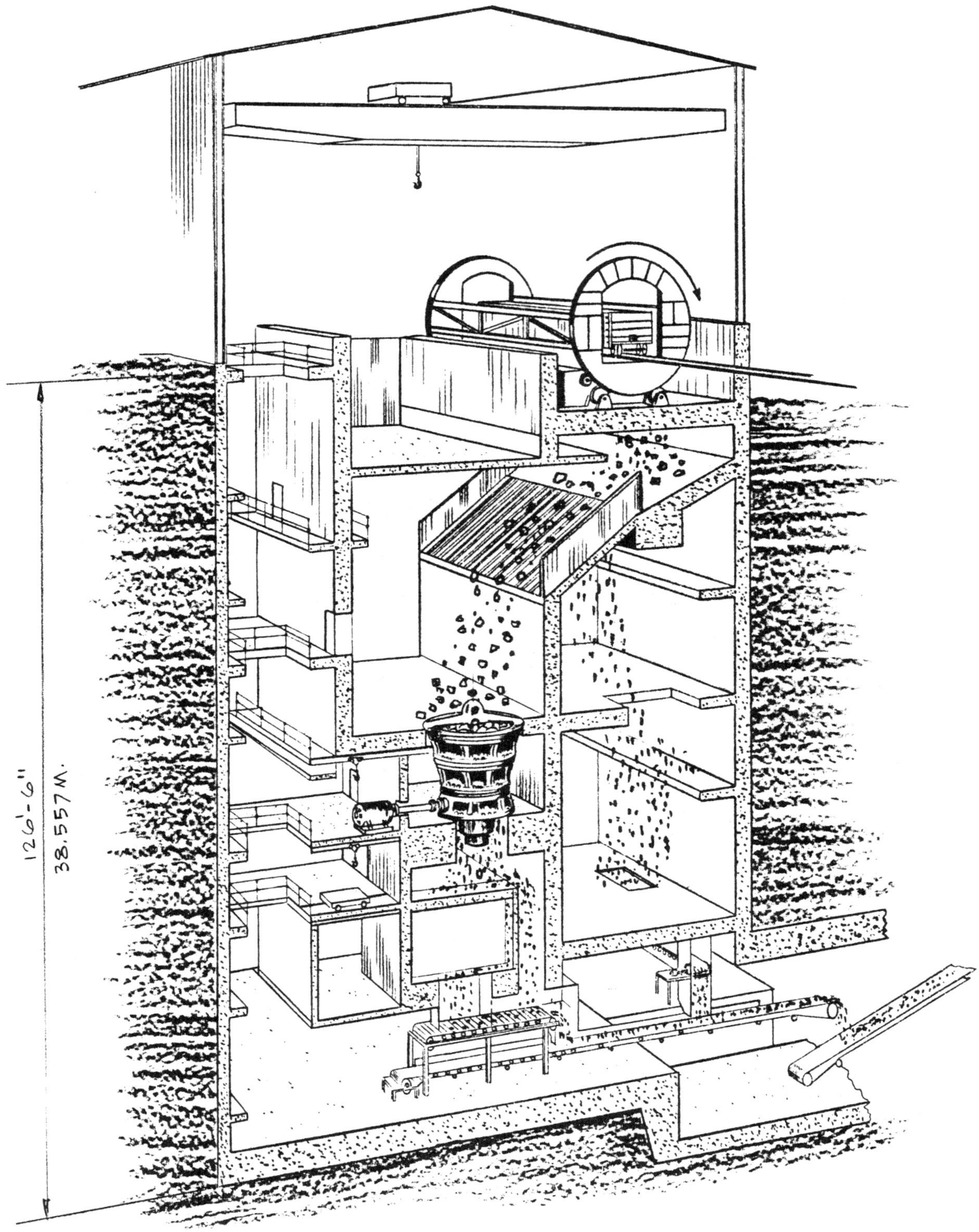

BONNEVILLE,
KENNECOTT COPPER CORP.
SALT LAKE CITY, UTAH

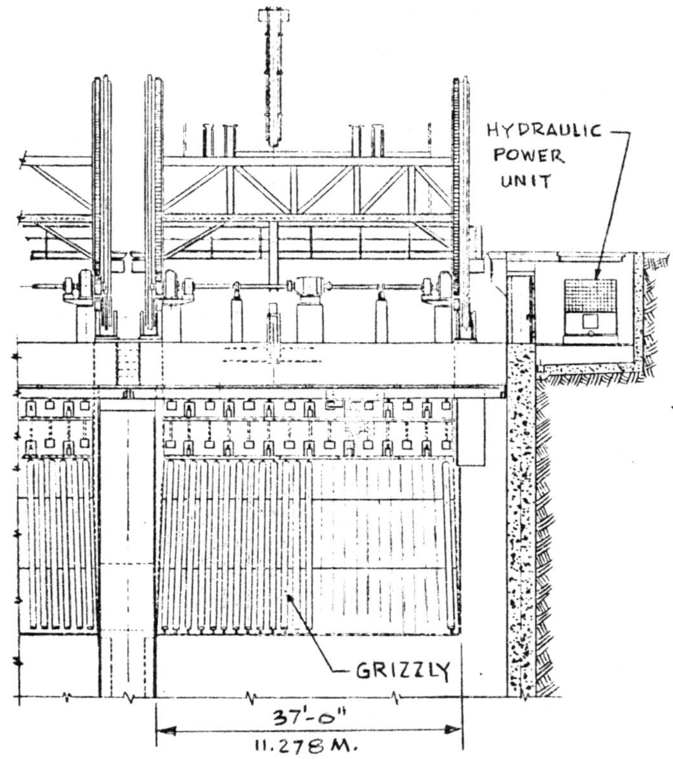

SECTION

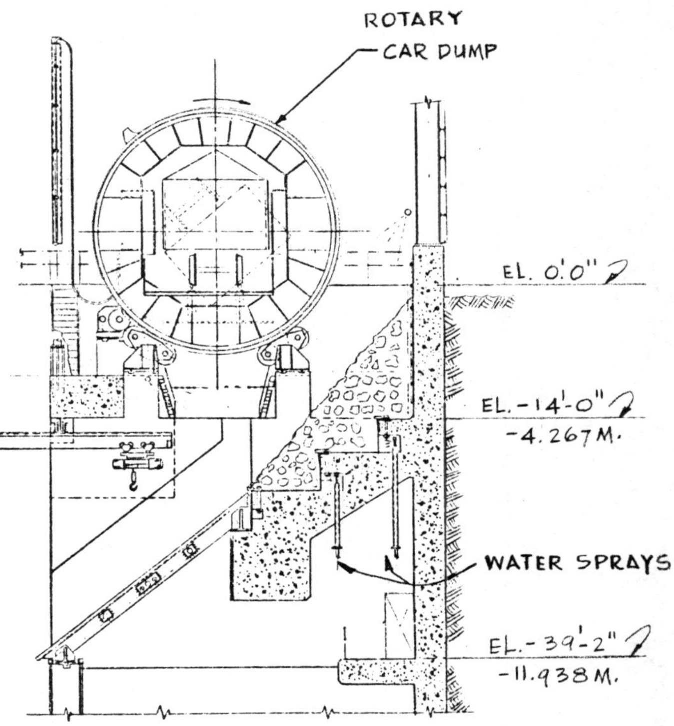

SECTION

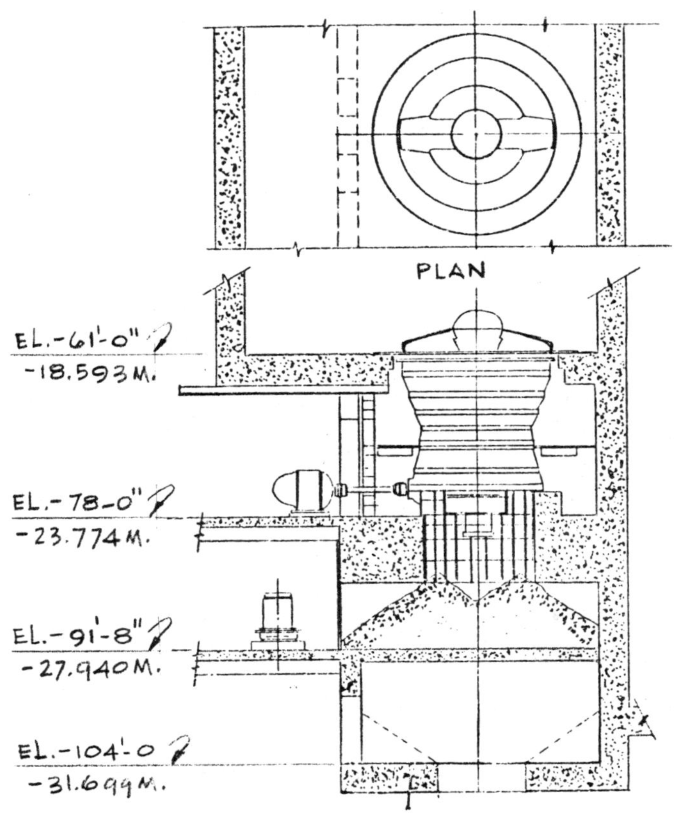

SECTION

BONNEVILLE,
KENNECOTT COPPER CORP.
SALT LAKE CITY, UTAH

KENNECOTT COPPER CORPORATION
RAY MINES DIVISION
RAY, ARIZONA

Prospectors were digging silver in the Ray mine area in 1873, and by 1880 high-grade copper ore was feeding a 30-ton copper furnace. In 1910, D. C. Jackling and his associates organized the Ray Consolidated Copper Company, known in mining circles as Ray Con, to exploit the Ray orebody. For 50 years, all mining was underground and theRay mine was known the world over for innovations in mining methods. It was the first to mine important tonnage (8,000 tpd) by block caving methods. The mill at Hayden, 26 miles (41.6 km) from the mine, kept pace with the new mining methods developed at Ray and in the early 1920's flotation replaced gravity concentration by the vanner machines. First mining by open pit methods began in 1950 and in January 1955 all underground mining was discontinued.

The new primary crushing plant, installed in early 1967, featured design simplicity with a low capital cost. The 54 in. gyratory crusher was installed within a total height of 61 ft (18.65 m) from the truck dump road bed to the floor beneath the feeder-conveyor under the surge pocket. The crusher is fed by 120-ton trucks dumping from one side only into a 200-ton hopper equipped with a horizontal splitter bar to protect the crusher spider.

The crushed ore pocket, of 120-ton live capacity underneath the crusher, discharges to a 0 to 200 fpm (61.2 mpm), variable-speed, 72 in. by 80 ft long (1.83 m by 24.4 m) sloped belt feeder, probably a first in the industry. Wood-picking stations are provided on either side of the belt so that old mine timbers and tramp iron can be removed from the crushed ore and the wood can be dropped into Dempster containers located below the belt. Counter-weighted finger gates control the bed depth on the belt at 14 in. (356 mm) and a low-level radiation unit automatically maintains a minimum of a 3 ft bed (900 mm) of material on the belt directly under the surge pocket. The conveyor is skirted for its full length and is approximately 60 in. (1.52 m) wide inside its metal skirt liners. Ten impact idlers at 12 in. (305 mm) spacing are provided under the loading pocket and remaining idlers are spaced at 18 in. (457 mm).

The belt is $1\frac{1}{4}$ in. (31.8 mm) thick; and, while it has occasionally been damaged by tramp material, belt life averages 11 million tons. When the belt is changed, the coiled new belt is lowered into position onto one of two belt stands mounted over the feeder, the old belt is cut, and one end is attached temporarily to the new belt. The other end of the old belt is then attached to the shaft of the second belt stand, which is turned by a permanently installed air motor. As the motor turns the shaft, the new belt is unrolled and the old belt coiled up. A V-type mechanical splice about 8 ft (2.44 m) long is used. Belt replacement requires about 8 hours.

A tramp iron magnet is provided over the head pulley of the belt feeder; that point, as well as a point on the stockpiling conveyor just downstream from its loading point, is observed by two television cameras, the receivers for which are mounted in the control room over the crusher.

The crushing plant design is a 53 ft by 47 ft 6 in. (16.15 m by 14.48 m) concrete box with the mainshaft stands and the maintenance area located at the eccentric removal level at an elevation of 35 ft 6 in. (10.22 m) below the dump level and at daylight, where truck access is provided. Pipes were installed at the back of the crushed ore pocket for storage of the eccentric removal rails, but they are no longer used.

An 8 ft by 10 ft (2.44 m by 3.05 m) control room is located at one corner of the dump pocket, 10 ft above the dump level. In the control room are two television sets which monitor the points mentioned previously. Alarm functions on the control panel are as follows:

Stockpile emergency low level
Stockpile 30-minute warning
Stockpile high level
Crusher oil pressure low
Crusher lube oil no flow
Crushed ore pocket high level
Crushed ore pocket high warning
Crushed ore pocket low warning
Crushed ore pocket low level
Transfer chute plugged
Stockpile belt slow speed
Stockpile belt skew

When the crushed ore pocket high or low warnings sound, the operator adjusts the speed of the reclaim feeder and, on signal from the crushed ore pocket high or low level, the reclaim feeder automatically stops. Clean crushed ore is transferred to a 54 in. (1.32 m) wide stockpile conveyor which discharges to a 4,000 live ton pile over a railroad load-out tunnel. Twenty-five thousand tons of ore are sent to the concentrator each day. Lately the capacity has been increased to 30,000 tons.

The cost of this crusher installation (not including the stockpile belt) was reported to be $1 million in 1967. This low-cost installation attests to a fine design.

KENNECOTT COPPER CORPORATION

Ray Mines Division

Ore type	Copper Porphyry
Concentrator capacity	24,500 st/day
Shovel size	7 yd (5.35 m^3)
Truck size	120 st
Primary crusher	54"x 74" Allis Chalmers Gyratory
Connected and consumed horsepower	500 - (not available)
Capacity of dump pocket	200 tons
Grizzly	Not used
Method of breaking or removing oversize	10 ton crane and I-R Rockbuster
Open side setting (product size)	Minus 8 in. (203 mm)
Eccentric throw	1 5/8 in. (41.4 mm)
Crusher drive	Direct
Capacity of crushed ore pocket	Approximately 120 tons live
Crushed ore feeder	72" belt, 0 to 200 fpm (1.83 m - 61 mpm)
Takeaway belt	54" at 450 fpm (1.37 m - 137 mpm)
Operating data:	
Shifts operated per week	21
Effective operating hours	Not available
Bulk density of ore	105-115 #/ cu ft (1680 - 1842 kg/m^3)
Nominal capacity	1,200 st/hour
Maximum capacity (peak)	2,500 st/hour
Effective availability, % of scheduled time	Not available
Backing for liners	Zinc
Mantle life	14 months
Concave life	18 months
Time to change mantle	16 hours
Time to change concaves	24 hours
Liner materials in dump pocket and crushed ore pocket	Manganese steel
Capacity of overhead service crane	60 ton main, 10 ton auxiliary
Coarse ore stockpile, after crusher	4,000 st live - 16,000 st total
Direct operating costs per ton	Not available
Overall height of installation	61 ft 0 in. (18.6 m)

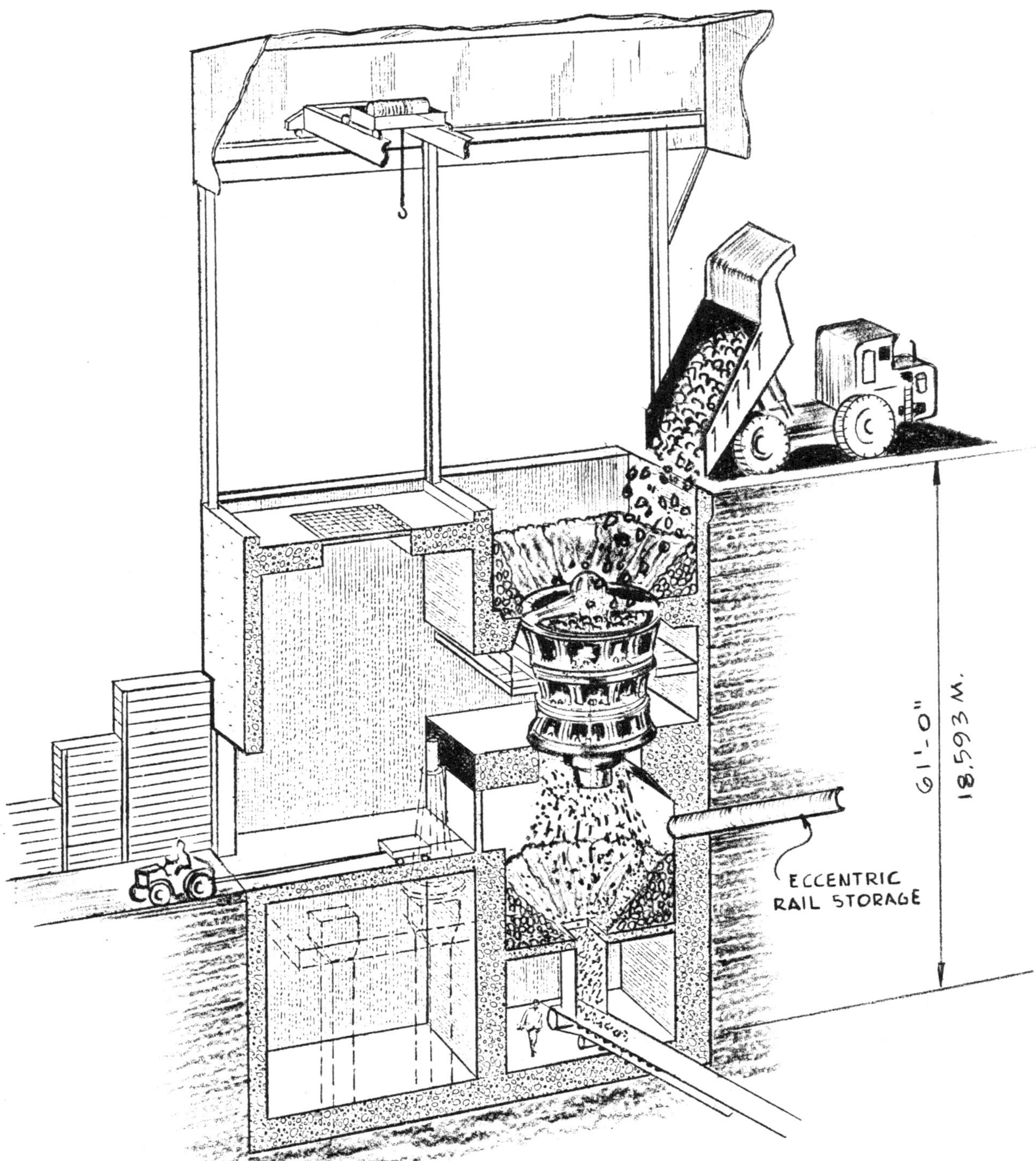

RAY MINES
KENNECOTT COPPER CORP.
RAY, ARIZONA

PALABORA MINING COMPANY
PHALABORWA, TRANSVAAL
REPUBLIC OF SOUTH AFRICA

The Palabora ore consists of two types, a carbonatite containing a low titanium magnetite, and a foskorite containing high titanium magnetite. The latter type also contains substantial quantities of phosphate. These two ore types are separately crushed and are kept separate through copper flotation and magnetic separation, for recovery of magnetite, with the low Ti magnetite being shipped as iron ore. High phosphate tailings are pumped to Foskor, a government-owned phosphate plant nearby. The Palabora plant was originally designed in 1966 for 33,000 tpd using two 54 in. by 74 in. Allis-Chalmers gyratories; two 7 ft (2.13 m) standards; four 7 ft shortheads; and six grinding lines, each using one 1,200 hp rod mill and two 1,200 hp ball mills. A fifth shorthead and one more grinding line were added later; and with flowsheet improvements, the capacity of the plant was increased to 59,500 stpd. A further plant expansion involving a third 54 in. primary and two 32 ft by 15 ft (9.75 by 4.6 m) 9,000 hp autogenous mills and another flotation section will bring its capacity to 87,500 stpd in 1977. These two fully autogenous mills are constructed entirely out of doors near the primary crusher, and the ground ore will be pumped a mile to the concentrator by a five-stage pumping system.

The Palabora ore is mined with 6, 10, and 12 7d (4.6, 7.6, and 9.2 cu m) shovels and hauled to the crushers in 110 and 165 st trucks. Sixty-five ton trucks were originally used, but the crusher dump pockets were designed for larger units. Dump pockets, crushers, feeders, and take-away belts are offset in order to keep the two ore types separate. Centerlines of the crushers are only 43 ft (13.1 m) apart; but, because of their being offset, maintenance room is ample. Initially, more carbonatite ore than foskorite was mined; but as the pit has deepened, this ratio has been reversed. The Palabora ore breaks in the pit with a large amount of oversize boulders; and, with the use of larger trucks, a splitter has been installed on one side of each dump pocket to protect the crusher spider. Smaller trucks dump on the opposite side. To accomodate the splitter, the pocket edge has been raised one meter and moved back one meter.

Dust is collected through a special vent shaft between the crushers and the mantle relining area; but even so, two high-pressure water spray headers with nozzles are installed under the edge of the pocket at each dumping position. While only one truck dumps at a time, sprays from all four headers are used to suppress dust. Any leakage of water through the seal around the upper bowl of the crusher is caught by a circular trough under the seal and directed to a sump pump. At this point, there is also a circular monorail and trolley which carries an air impact wrench on a counter-weighted beam. Loosening or tightening the bolts on the upper bowl are greatly eased by this novel arrangement.

When water contacts the Palabora ore, it becomes extremely corrosive even though the pH is in the range of 8. Consequently, reclaimed water lines must be stainless steel or rubber lined; and all slurry lines, flotation cells, etc., also are rubber lined. Embedded metal which may become wet from washdown water is stainless steel, and even conveyor decking where dust collects must be stainless. This corrosive water has probably been a partial cause of primary crusher main shaft breakage. Five of these shafts have been broken, four of them at the radius in the shaft between the head center and the head nut, and one a few inches above this point. It was determined that water and dust from the dump pocket dust suppression sprays entered through the dowel holes in the head nut and caused corrosion fatigue cracking. This situation was aggravated by the shaft being put in enormous tension by the flowing of the manganese steel upper mantle liner. Crushing of the great number of unusually large rocks in the upper part of the crusher bowl was determined as the cause. The Palabora ore is not particularly abrasive, but it is tough and mantle liners last six to eight months. Filling of the annular space at the head nut with heavy grease and altering the upper liner changeout times have been adopted to prevent the occurrence of more main shaft breakages.

Maintenance crews are particularly well organized, as exhibited by changeout times for mantles and concaves of 6 and 21 hours respectively. Both mantle liners and concaves are zinced. Crusher maintenance is done with an overhead crane having hoists of 75 and 15 tons capacity. Operating availability is a phenomenal 94% of the scheduled operating time of 18 shifts per week. Mining on Sunday is not allowed in South Africa.

The crushed ore pockets have live capacities of 150 st each, and the ore is transferred to the 60 in. (1.52 m) takeaway belts by 72 in. (1.83 m) wide apron feeders operating at 33 fpm (10.1 mpm). Takeaway belts transfer the ore to separate stockpiles, which in turn feed the fine crushing plant.

The electrical room is below grade beneath the repair area. Concrete usage was 8,566 yd (6550 cu m) for this twin crusher enclosure which is only 74 ft 8 in. (22.76 m) in depth from the top of the dump pocket to the top of concrete of the takeaway conveyor floor. Operating costs of these two crushers in May 1976 were less than five cents per ton.

PALABORA MINING COMPANY

Phalaborwa, Transvaal, Republic of South Africa

Ore type	Carbonatite, Foskorite and Magnetite; 127#/cu ft (2035 kg/m^3)
Concentrator capacity	59,500 st
Shovel sizes	6, 10, and 12 yd (4.6, 7.6, and 9.2 m^3)
Truck sizes	110 and 165 st
Primary crusher	Two 54" x 74" Allis Chalmers Gyratories
Connected and consumed horsepower	500 each - 1118 total, includes auxiliaries
Capacity of dump pockets	660 st filled
Grizzly	Not used
Method of breaking or removing oversize	Removal by overhead crane - rock hook
Open side setting (product size)	8.1 in. (206 mm)
Eccentric throw	1 5/8 in. (41.5 mm)
Crusher drive	Direct
Capacity of crushed ore pocket	150 st each
Crushed ore feeder	72" apron feeder at 33 fpm (1.83 m at 10.1 mpm)
Operating data:	
Shifts operated per week	18
Effective operating hours	8 per shift
Nominal capacity	1,265 st/hour each
Maximum capacity (peak)	2,175 st/hour each
Effective availability	94% of scheduled time overall
Backing for liners	Zinc
Mantle life	6 to 8 months
Concave life	12,100,000 st
Time to change mantle	4 hours
Time to change concaves	21 hours
Liner materials in dump pocket and crushed ore pocket	Manganese steel
Capacity of overhead service crane	75 ton main, 15 ton auxiliary
Coarse ore stockpile after crusher	38,500 st live
Direct operating costs per ton (includes labor power, supplies, maintenance)	$0.0483/st (May 1976)
Concrete	8,566 cu yds (6550 m^3)
Reinforcing bar	399 st
Embedded steel	8.8 st
Structural steel	287 st
Miscellaneous steel	169 st
Overall height of installation	74' 8" (22.76 m)

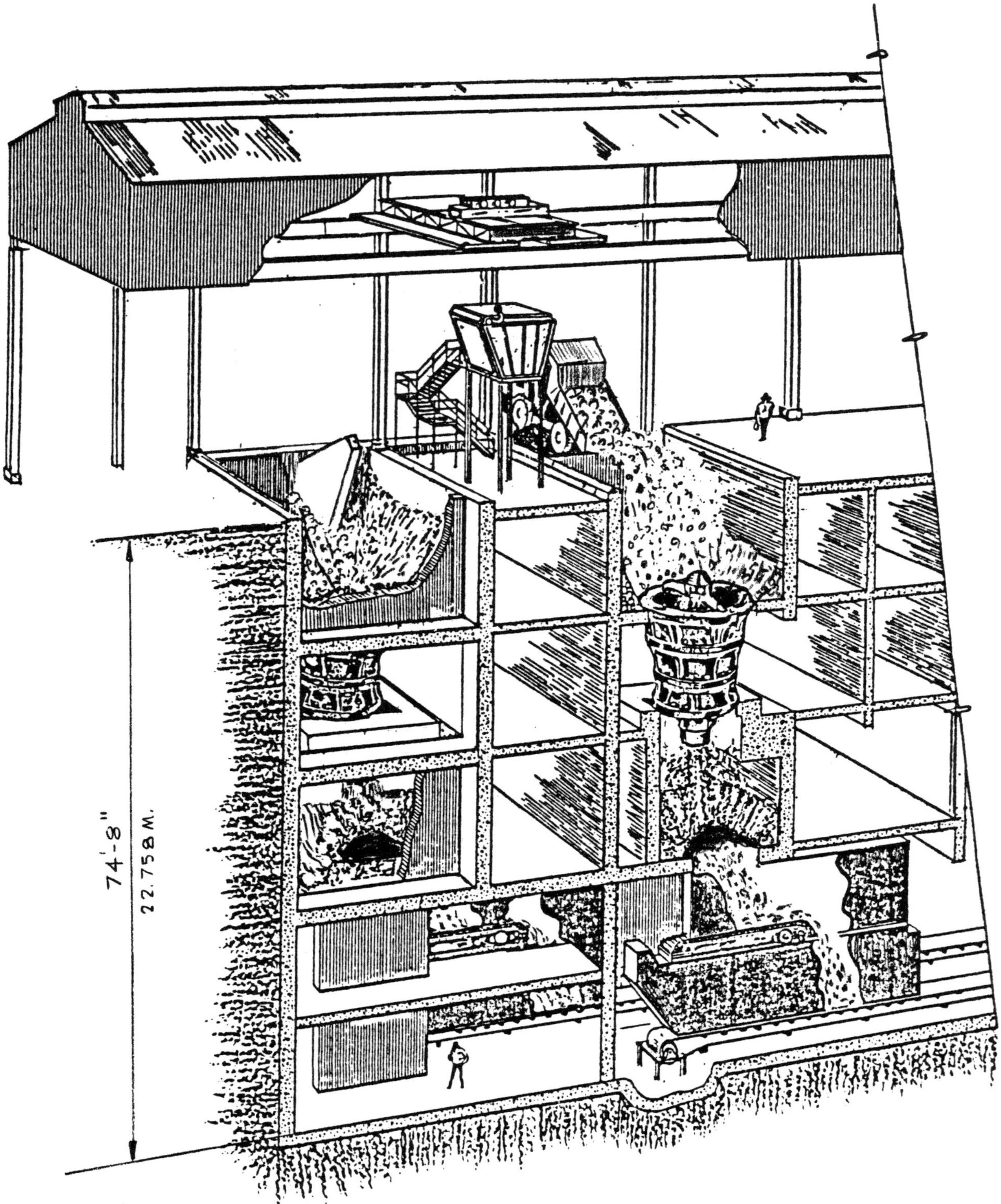

PALABORA MINING CO., LTD
PHALABORWA, TRANSVAAL
SO. AFRICA

PALABORA FIG. 8

PALABORA MINING COMPANY
PHALABORWA, TRANSVAAL
REPUBLIC OF SOUTH AFRICA

As described in the section on the Palabora twin crushers, the Palabora ore consists of two types: a carbonatite containing a low titanium magnetite, and a foskorite containing high titanium magnetite. As the Palabora pit was widened and deepened, more and more of the foskorite had to be mined. The 25,000 mt expansion of the Palabora plant which went into operation in early 1977 therefore takes its feed from the foskorite section of the orebody. For the expansion, a third 54 in. by 74 in. gyratory crusher was installed to crush feed for two 32 ft by 15 ft (9.75 by 4.6 m), 9000 hp (6710 kw) autogenous single-stage grinding mills. These units are located near the two existing primary crushers, and the ground ore is pumped 5277 ft (1640 m) to the expanded flotation plant by five stages of centrifugal pumps.

The new gyratory is installed in a unique Figure-8 shaped enclosure with the crusher in the smaller 27 ft 11 in. (8.5 m) diameter cylinder and the maintenance area in the 36 ft 1 in. (11 m) diameter cylinder. Cylinder walls are only 23.6 in. (600 mm) thick because their circular design permits all the concrete to be in compression. A rectangular dump pocket which sits on the smaller cylinder contains two manganese steel faced splitters to protect the crusher spider and countersunk water spray headers under the dumping positions for dust suppression. The dump pocket holds 330 st or 300 mt and is also lined with manganese steel.

The crusher is mounted on a reinforced concrete diaphragm over the 363 st (400 mt) crushed ore pocket. In the back of the pocket is plastic pipe. Radioactive level control sources are lowered into this pipe from the crusher support floor. The pipe is protected by concrete covered with steel castings. Level control detectors are mounted on the front wall of the pocket. Liner plates on the crushed ore pocket walls are removed and installed by the use of a circular monorail and trolley mounted on the pocket ceiling.

The door in the front of the crushed ore pocket is hinged at the top rather than at the sides and is raised by an electric winch. For removal of the eccentric, rails are installed in the pocket and a unique hydraulically operated cart is moved under the eccentric. A platform is raised to receive the eccentric and lowered so that it may be removed.

Crushed ore is removed from the pocket by a hydraulically powered, 8 ft wide by 20 ft long (2.44 m by 6.1 m) belt feeder. The feeder is similar to the one used at Similkameen except the belt is flat rather than being turned up at the edges. A vee-shaped opening, relieved both horizontally and vertically, is used above the feeder to prevent belt and skirt wear and to permit drawdown of the ore over the full length of the opening in the bottom of the crushed ore pocket. At the top of the feeder skirts are holes for spile bars, and a small winch mounted on a monorail is used for spile bar removal. Another monorail, this one circular, is mounted on the underside of the dump pocket and around the top bowl section of the crusher. Hanging from a trolley on this monorail is a counterweighted air impact wrench which facilitates removal and installation of the upper ring bolts. Other monorails are installed over the feeder drive and the crusher jackshaft.

In the maintenance cylinder are a stairway, a man elevator, and the usual hatchways for hoisting maintenance items and equipment. Also in this area is the crusher lubrication room and a unique hydraulic accumulator for furnishing oil to the crusher bearings during the rundown of the crusher due to a power failure. The takeaway conveyor is sloped at its loading point and does not extend completely under the belt feeder since few ore fines stick to the rubber belt the way they would to an apron feeder. A belt scraper is used on the feeder belt.

An electrical room as well as a wet dust collector are located at ground level. Dust is therefore drawn upward from the belt feeder transfer point, slurried in the collector, and pumped to the grinding area. Over the crusher is an 80/20 ton crane for mainshaft removal. Because the new (third) primary is near the two other primaries, the mainshaft is transported there for replacement of the mantle liner. Liners are backed with zinc.

By using the Figure 8 concept for the crusher enclosure, a substantial saving in steel and concrete was realized. Walls of a conventional enclosure of a similar depth and with straight walls are generally 4 ft (1.22 m) thick, whereas the walls of the Figure 8 enclosure are only 23.6 in. or 600 mm thick. The Figure 8 design is patented by Bechtel Corporation.

PALABORA MINING COMPANY

Phalaborwa, Transvaal, Republic of South Africa

Ore type	Carbonatite, foskorite, and magnetite 127#/cu ft (2035 kg/m^3)
Concentrator capacity	27,560 stpd (25,000 mtpd)
Shovel sizes	6, 10, and 12 yd (4.6, 7.6, and 9.2 m^3)
Truck sizes	110 and 165 st
Primary crusher	54 in. by 74 in. Allis Chalmers gyratory
Connected horsepower	500 (373 kw)
Grizzly	Not used
Method of breaking or removing oversize	Removal by overhead crane - rock hook
Open side setting (product size)	8.1 in. (206 mm)
Eccentric throw	1 5/8 in. (41.5 mm)
Crusher drive	Direct
Capacity of dump pocket	330 st (300 mt)
Capacity of crushed ore pocket	441 st (400 mt)
Crushed ore feeder	96 in. by 20 ft (2.44 m by 6.1 m) rubber belt
Takeaway belt	60 in. (1.52 m)
Operating data:	
Shifts operated per week	18
Effective operating hours	8 per shift
Nominal capacity	1265 st
Maximum capacity (peak)	2175 st
Effective availability	94% of scheduled time over all
Backing for liners	Zinc
Mantle life	6 to 8 months
Concave life	12,000,000 st (10,900,000 mt)
Time to change mantle	4 hours
Time to change concaves	21 hours
Liner materials in dump pocket and crushed ore pocket	Manganese steel - A.R. plate
Capacity of overhead service crane	80 - 20 mt
Coarse ore stockpile	Two @ 30,900 st - 28,000 mt live
Concrete	7680 cu yd (5870 m^3)
Reinforcing bar	480 st (443 mt)
Overall height of installation	101 ft 1 in. (30.8 m)
Date of completion	December 1976

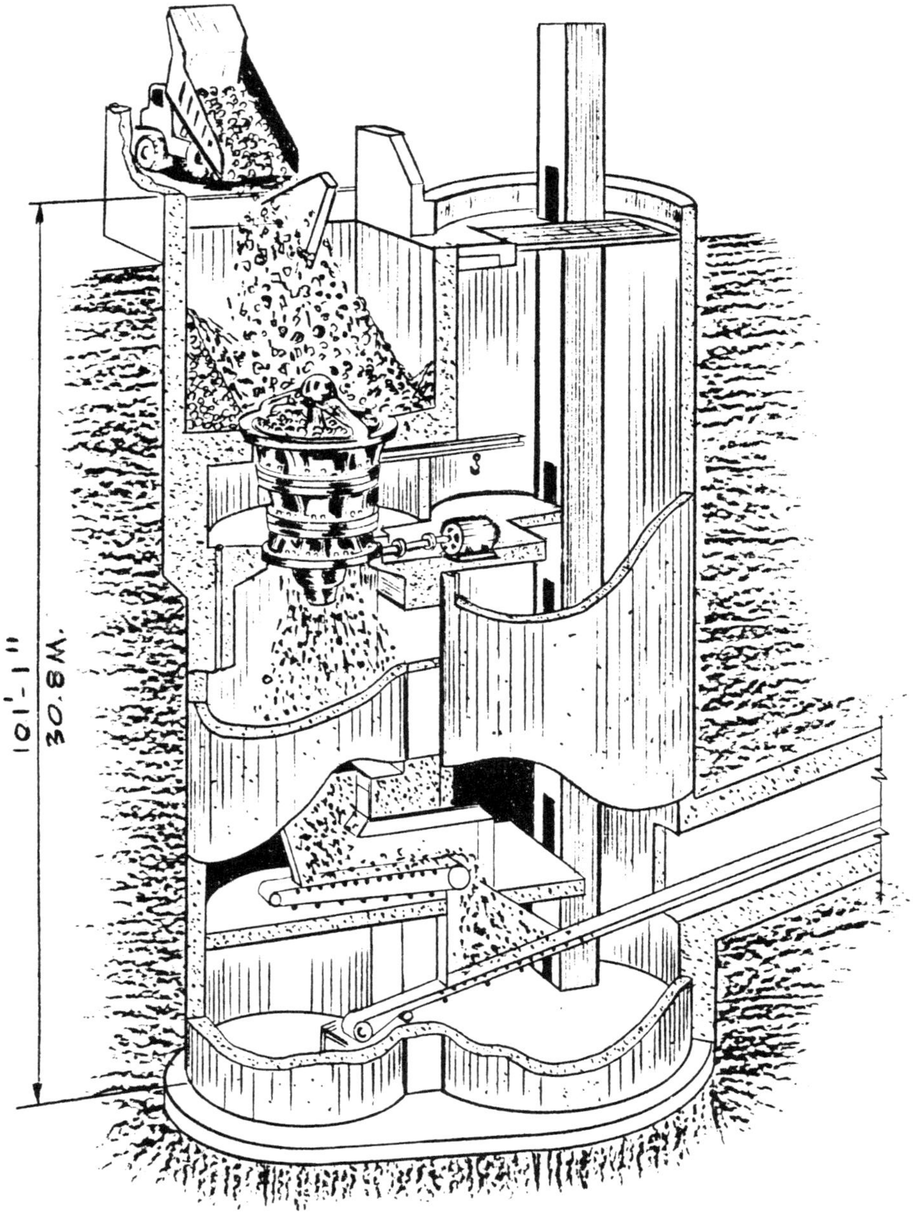

SECTION

THE FIGURE 8 DESIGN
OF THIS CRUSHER ENCLOSURE
HAS BEEN PATENTED BY
BECHTEL CORPORATION.

PALABORA MINING CO. LTD.
PHALABORWA, TRANSVAAL
SO. AFRICA

SIMILKAMEEN MINING COMPANY, LIMITED
PRINCETON, BRITISH COLUMBIA

The Similkameen mine is located about 100 miles east of Vancouver, British Columbia, and ten miles west of Princeton, where the mine personnel live. Princeton was the first town in the British Columbia interior, originally the home of Cree and Blackfoot Indians and later settled by men of the Hudson's Bay Company, miners, cattlemen, and others looking for gold. After gold, there were coal mines and then copper mines.

The Granby Consolidated Mining Company, Limited, operated the Copper Mountain ore bodies from 1926 to 1930 and from 1937 to 1957, mining 34 million tons of about 1% copper ore, mainly from underground. In 1966, Newmont Mining Corporation acquired the Ingerbelle property on the west side of the Similkameen River. Newmont then purchased the assets of Granby and formed Similkameen Mining Company to develop and operate the Ingerbelle and Copper Mountain ore bodies on both sides of the Similkameen River canyon.

The Ingerbelle and Copper Mountain ore bodies lie within rocks composed of andesite tuffs, agglomerates, and argillites. Chalcopyrite mineralization occurs as a dissemination. Original ore reserves were estimated at approximately 75 million tons grading 0.52% copper.

Mining is carried out on a three-shift, seven-day-week schedule at a rate of about 60,000 tons of ore and waste per day with four 10 cu yd P & H 1900 shovels which mine and load a fleet of 100-ton Lectra Haul trucks.

Among the unique features of the design and construction of the Similkameen project was the construction of 4.5 miles (7.24 km) of new Trans-Provincial No. 3 highway by the company to the specifications of the British Columbia Department of Highways. The old highway traversed the Ingerbelle pit area. Along the relocated highway is a 4,000 ft (1219 m) long berm which conceals the mine pit. This area has been replanted and now looks like the original forest scenery. Another interesting feature is that the tailings from the concentrator flow by gravity over a 1000 ft (348 m) suspension bridge, which crosses the Similkameen River canyon then passes through a tunnel to the dam site, a distance of 59,000 ft (17,980 m) to the cyclone station. When the Copper Mountain ore bodies are mined, the ore will be transported by belt conveyor on a second suspension bridge across the canyon to the Ingerbelle concentrator.

After being loaded by the shovels into the 100-ton trucks, the ore is transported to the mill site and dumped into a single-side dump, 54 in. by 74 in. Allis-Chalmers gyratory crusher, where it is reduced to minus 9 in. (239 mm). The crushed ore is conveyed to a luffing stacker where it is spread over a 45,000 live ton storage pile. From the pile, Hydrastroke feeders draw ore to feed the three 8,000 hp, 32 ft diameter by 14 ft long (9.75 m by 4.26 m) semi-autogenous mills.

Newmont and Bechtel engineers agreed a considerable capital cost reduction could be made, without increasing operating costs or the loss of efficiency, by departing from standard primary crushing plant design practices. Consequently, the superstructure and 50-ton overhead bridge crane were deleted. For handling the main shaft, an 82-ton mobile crane was substituted for the 40-ton crane required as part of the open pit equipment; and for handling the oversize rock, a Ramey hydraulic grapple was added. This resulted in a cost reduction of approximately $250,000 (1970). Confirmation that the mobile crane has worked well was demonstrated by the relining times for the mantle and concaves: 24 and 72 hours respectively for the first relining with an inexperienced crew. The dump pocket is lined with 1 in. (25.4 mm) thick manganese steel and the crushed ore pocket with AR plate. Originally, the dump pocket was equipped with a splitter beam to protect the crusher spider; but it was damaged so often it was removed. Dumping ore directly on the spider has caused no problems.

A further cost reduction of $175,000 was made through the substitution of a 96 in. by 50 ft long (2.44 m by 15.24 m) belt feeder in place of the more conventional apron feeder under the 165-ton crushed ore pocket. The extra-long feeder permitted the downward sloped takeaway conveyor to be straight instead of having a reverse vertical curve in it. The feeder uses picking-type idlers for supporting the $1\frac{1}{2}$ in. (38 mm) thick belt, which is still in service (January 1977) after handling 26 million tons. Because of the success of this feeder, similar units have been installed at both Gibraltar in British Columbia and Palabora in South Africa. The Similkameen belt feeder was adapted from a crushed ore belt feeder installed at Ray. It is interesting to note, however, that the installed total height of the Similkameen crusher enclosure is only 59 ft (18 m), two feet (0.6 m) less than that at Ray. It is perhaps the lowest installed height for a 54 in. gyratory.

The very shallow dump pocket, 9 ft 8 3/4 in. (2.97 m) combined with the 165-ton crushed ore pocket and the belt feeder resulted in an installation which used only 2,200 cu yd (1682 cu m) of concrete, including the wing walls holding the embankment. The total installation cost (1972) was approximately $2 million, and the operating cost in the first five months of 1975 was $0.05 per ton.

SIMILKAMEEN MINING COMPANY
Princeton, British Columbia

Ore type	Copper - Pyrometasomatic, 175#/cu ft in place (2804 kg/m^3)
Concentrator capacity	22,500 st/day
Shovel size	10 yd (7.6 m^3)
Truck size	100 st - 60 cu yd boxes (46 m^3)
Primary crusher	54" x 74" Allis Chalmers Gyratory
Connected and consumed power	600 hp - 0.7 kwh/st
Capacity of dump pocket	300 st (approximate)
Method of breaking or removing oversize	Hydraulic rock picker with rock grapple
Open side setting (product size)	8 1/2 in. (216 mm)
Eccentric throw	1 5/8 in. (41/5 mm)
Crusher drive	Direct
Capacity of crushed ore pocket	165 st
Crushed ore feeder	96" x 50' long variable speed conveyor belt (2.44 m by 15.24 m)
Takeaway belt	48" @ 450 fpm (1.22 m @ 137 mpm)
Operating data:	
Shifts operated per week	21
Effective operating hours	8 per shift
Nominal capacity	1500 st/hour
Maximum capacity (peak)	2500 st/hour
Effective availability	Not applicable - more time available than required
Backing for liners	Concaves: epoxy; Mantle: zinc
Mantle life	8,200,000 st
Concave life	12,700,000 st
Time to change mantle	Two 12-hour shifts
Time to change concaves	Six 12-hour shifts (changed only once)
Liner materials in dump pocket and crushed ore pocket	AR Plate 1 inch and 1 1/2 inch (25.4 - 38 mm)
Capacity of service crane	82 ton Link Belt mobile unit
Coarse ore stockpile (after crusher)	40,000 st - live, 140,000 st - total
Direct operating costs per ton (includes labor, power, supplies, maintenance)	$0.05/ton (first five months of 1975)
Concrete	2,200 cu yd (1682 m^3)
Reinforcing bar	105 tons
Structural steel	23 tons
Miscellaneous steel	58 tons
Overall height of installation	59 ft 0 in. (18 m)
Constructed cost (excludes crushed ore conveyor)	ca $1,400,000 (September 1972)

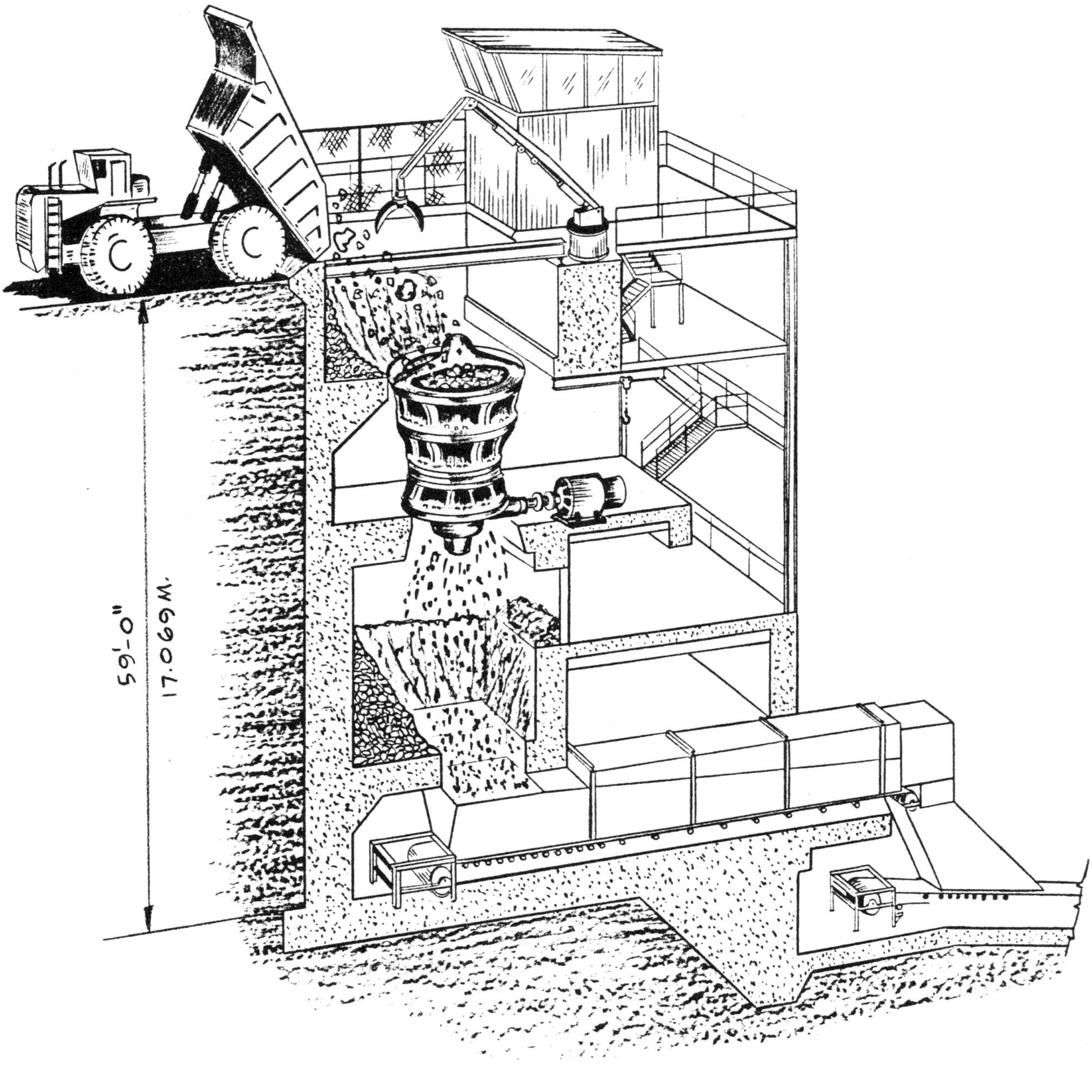

SIMILKAMEEN MINING CO.
PRINCETON, BRITISH COLUMBIA

60 INCH GYRATORY CRUSHERS

MINE	Anaconda	Cities Service	Duval	Iron Ore Co. of Canada
Location	Butte, Montana	Pinto Valley, Arizona	Sierrita, Arizona	Labrador City, Newfoundland
Ore type	Copper porphyry	Copper porphyry	Copper porphyry	Iron ore, hematite
Crusher size	60 in. by 89 in.	60 in by 89 in.	(2) 60 in. by 89 in.	(2) 60 in. by 89 in.
Capacity dump pocket	(2) 400 t each	300 st	250 st	300 st
Crusher feeder	(2) 84 in. by 384 vibrating	None	None	None
Grizzly	4 in.	None	None	None
Removing oversize	Crane, rock hook	Hydraulic rock breaker	Crane, rock hook	Drill and blast
Product size	5.5 in.	8 in.	6 in.	7 in.
Nominal capacity	5000 st/hr	3500 t/hr	2500 st/hr each	4000 lt each
Capacity crushed ore pocket	200 t	360 st	300 t	160 lt
Crushed ore feeder	72 in. by 192 in. vibrating	84 in. by 20 ft apron	84 in. apron	72 in. by 14 ft apron
Takeaway belt	60 in. @ 750 fpm	60 in. @ 530 fpm	60 in. @ 630 fpm	60 in. @ 650 fpm
Capacity service crane	75 t/20 t auxilliary	80 t/20	70 t/20 t	75 t/10 t
Direct operating cost/ton	$0.10 incl. wood 7/75	$0.038 June 1976	$0.042 Jan-June 1975	N. A.
Concrete used	12,800 cu yd	4,874 cu yd	8,000 cu yd	
Overall height installed	122 ft 0 in.	86 ft 6 in.	89 ft 0 in.	97 ft 6 in.
Installation cost	$8,300,000 1971	ca $5,250,000 - 12/74*	N. A.	N. A.
Comment	High cost due to feeder, grizzly, and wood picking			

*cost excludes crushed ore tunnel and conveyor

60 INCH GYRATORY CRUSHERS

MINE	Lornex	Reserve Mining	Southern Peru
Location	Logan Lake, B. C.	Babbitt, Minnesota	Cuajone, Peru
Ore type	Copper porphyry	Iron ore, taconite	Copper porphyry
Crusher size	60 in. by 89 in.	60 in. by 100 in.	60 in. by 89 in.
Capacity dump pocket	1000 st filled	200 lt	(2) Railroad cars @ 50 cu yd
Crusher feeder	None	None	None
Grizzly	None	None	16 ft 9 in. by 2-22 ft 4 in.
Removing oversize	Hydraulic hammer	Crane, rock hook	Barco manipulator
Product size	10 in.	$8\frac{1}{2}$ in.	8 in.
Nominal capacity	4000 st/hr	3000 lt/hr	4000 st/hr
Capacity crushed ore pocket	300 st	150 lt	5000 grizzly + 900 t
Crushed ore feeder	96 in. by 23 ft apron	Gravity to secondaries	(2) 84 in. aprons
Takeaway belt	60 in. @ 513 fpm	60 in. @ 380 fpm	60 in. @ 650 fpm
Capacity service crane	75 t		80 t/20 t
Direct operating cost/ton	N. A.	$0.175 inc. secondary crusher	N. A.
Concrete used	6255 cu yd	11,000 cu yd	8,400 cu yd
Reinforcing and other steel	922 t	1600 t	1560 t
Overall height installed	85 ft 3 in.	165 ft 0 in.	124 ft 10 in.
Installation cost	N. A.	N. A.	$6,400,000 not including frt. grizzly and railroad dump
Comment			

THE ANACONDA COMPANY
BUTTE, MONTANA

"The richest hill on earth" had its start as a placer gold camp in 1864. The gold was quickly worked out but then silver was discovered in the copper ores in Butte Hill. After ten years as a silver camp, rich copper ore was found in 1881 and the future of Butte was assured. In the early 1960's, the Anaconda Company began the shift from underground mining at Butte to open pit mining of the enormous, low-grade copper deposit surrounding the old high-grade veins which had been worked for over eighty years.

Their first primary crusher was a 66 in. by 84 in. (168 m by 2.13 m) jaw crusher. The pit was expanded at such a rate, however, that its location became endangered. Additionally, the new Clyde Weed concentrator had been built at Butte; and in 1969 construction of a new gyratory installation was started.

The primary crusher at the Clyde Weed concentrator is one of the most unusual primary crushers ever put into operation. At times, the ore from the Berkeley pit contains large amounts of timber, pipe, rails, and so forth left in the ore body from the old underground mining operations; and the crusher installation had to be designed so this tramp material could be readily removed ahead of the crusher. Ore is delivered by 100- and 150-ton trucks to two side-by-side dump pockets, each with a capacity of 400 tons. Originally, a hydraulic grapple was installed on the dividing wall between the dump pockets; but it was found that oversize boulders large enough to block the discharge opening at the bottom of each hopper were rapidly covered with other material, and the grapple was unable to reach and remove the large boulders causing the blockage. In addition, the grapple was located in a quite vulnerable position; and it sustained damage from being struck by rock dumped from the trucks. Oversize boulders are now blasted. Underneath each pocket is a vibrating grizzly feeder with the grizzly bars set at 4 in. (102 mm). These feeder-grizzlies are 84 in. (2.13 m) wide and 32 ft (9.75 m) long, with grizzly sections 16 ft (4.88 m) long.

Approximately 75% of the ore passes through the grizzlies and grizzly oversize then drops onto two parallel 6 ft (1.83 m) wide picking belts. The belts are 3 in. (76 mm) thick and rest on idlers supported by air-filled rubber shock absorbers. At the side of each picking belt is a hydraulic grapple which is used to remove timbers too large to be removed by hand. At times, four men are used to remove timber and tramp iron. Timbers are dropped onto a transverse belt conveyor which discharges into a truck stationed outside the primary crusher building, and tramp iron is put into tote boxes. The picking belts discharge their ore on either side of the spider of a 60 in. by 89 in. Traylor gyratory crusher.

In the Clyde Weed concentrator only secondary crushing to $1\frac{1}{4}$ in. (32 mm) is done before rod milling. Originally, at the primary crusher, both the grizzlies and the gyratory were set at 7 in. (178 mm) but when hard ore was encountered the three 7 ft (2. 15 m) standard crushers became overloaded. Feed to the standards then had to be reduced to maintain the proper size for the rod mill feed. By decreasing the grizzly and primary crusher settings to 4 in. (102 mm), the operators found that the resulting smaller-sized feed to the secondary crushers permitted a 10 to 14% increase in crushing rate of hard ore, thus nearly maintaining the 52, 500 tpd throughput in the concentrator.

Under the primary crusher, the grizzly undersize is transferred to the takeaway belt by a 5 ft by 15 ft (1.52 m by 4. 57 m) apron feeder, while the crushed ore is fed onto this same belt by a 6 ft by 16 ft (1. 83 m by 4. 88 m) vibrating feeder. Nominal capacity of the crusher installation is 5, 000 short tons per hour and maximum peak capacity is 6, 000 tons per hour.

Access to the crusher motor and the eccentric is through hatches on either side of the crusher. Rails for carrying the eccentric removal cart are permanently mounted in the small ore pass immediately beneath the crusher.

Because some tramp iron invariably passed the vibrating grizzly, the takeaway belt under the crusher was often subject to damage and long tears in the belt were a fairly common occurrence. Anaconda personnel developed a unique belt rip detection device which has solved this problem. Each of the impact idlers at the transfer points from the two feeders to the takeaway conveyor is hinged on the leading edge of the idler bracket. The trailing edge is held down by a heavy spring. When a piece of tramp metal penetrates the belt and jams against the idler, the idler moves forward slightly, and this movement is detected by a proximity switch on the idler bracket. This then stops the conveyor before anything more than minor damage to the belt can occur.

An extensive ramp made of mine waste gives access to the dump pockets by mine trucks. Another unique feature is truck access at the level of the feed pocket above the crusher. Because dust suppression water sprays cannot be used in the truck dump pocket, due to the cold weather at Butte, air is ducted through a slot in the back wall of each dump pocket to a 10, 000 cum (283 cu mpm) baghouse. A similar baghouse collects dust from within the crusher enclosure. Collected dust is pugged with water and transferred by a screw conveyor to the inclined crushed ore takeaway belt. Construction of this crushing plant began in October of 1969 and was completed in February of 1971. A total of 12,800 cu yd (9787 cu m) of concrete were used in the crusher enclosure, which included 1,100 cu yd (840 cu m) in the 4 ft thick by 64 ft wide by 80 ft long (1.22 m by 19.5 m by 24.4 m) supporting slab that forms the bottom floor of the crushing plant. In addition, a total of 1,100 tons of reinforcing steel were utilized. The entire crushing system, which includes a 60 in. wide by 950 ft long (1.52 m by 290 m) takeaway conveyor, had a total cost of slightly over $8 million, of which dust collection equipment accounted for $600,000 and the gyratory crusher itself $515,000.

THE ANACONDA COMPANY

Butte, Montana

Ore type	Copper - Porphyry, 174#/cu ft in place (2786 kg/m^3)
Concentrator capacity	52,000 tons per day
Shovel sizes	6 yd and 15 yd (4.6 and 11.5 m^3)
Truck sizes	100 and 150 st
Primary crusher	Traylor 60" x 89" Gyratory
Connected and consumed horsepower	500 - 300
Capacity of dump pocket	Two pockets, 400 tons each
Crusher feeder	Two 84" x 384" mechanical vibrating grizzly feeders (2.13 m x 9.75 m)
Length and width of grizzly section	16' x 84" (4.87 m x 2.13 m)
Spacing of bars	4 in. (102 mm)
Estimated % ore through grizzly	75%
Method of breaking or removing oversize	Removal using 20-ton overhead crane and sling
Open side setting (product size)	5.5 in. (140 mm)
Eccentric throw	1.5 in. (38 mm)
Crusher drive	Vee belts
Capacity of crushed ore pocket	200 tons
Crushed ore feeder	72" x 192" mechanical vibrating (1.83 m x 4.88 m)
Grizzly undersize feeder	60 in. by 15 ft apron, variable speed to 40 fpm (1.52 m x 4.6 m @ 12.2 mpm)
Takeaway belt	60" x 950' @ 750 fpm (1.52 m x 290 m @ 229 mpm)
Operating data:	
Shifts operated per week	16
Effective operating hours	6.5 per shift
Nominal capacity	5,000 st/hour
Maximum capacity (peak)	6,000 st/hour
Effective availability	95% of scheduled time
Backing for liners	Zinc
Mantle life	4.8 million tons
Concave life	9.7 million tons
Time to change mantle	6 hours
Time to change concaves	48 hours
Liner materials in dump pocket and crushed ore pocket	14% manganese
Capacity of overhead service crane	75 ton w/20-ton auxiliary
Coarse ore stockpile (after crusher)	25,000 tons live - 55,000 tons total
Direct operating costs per ton (includes labor, power, supplies, maintenance)	$0.10/ton (crushing, wood picking and conveying, July 1975)
Concrete	12,800 cu yd (9787 m^3)
Reinforcing bar	1,100 tons
Cost including 950' crushed ore conveyor	$8,300,000 (February 1971)
Overall height of installation	122' 0" (36.57 m)

THE ANACONDA CO.
BUTTE, MONTANA

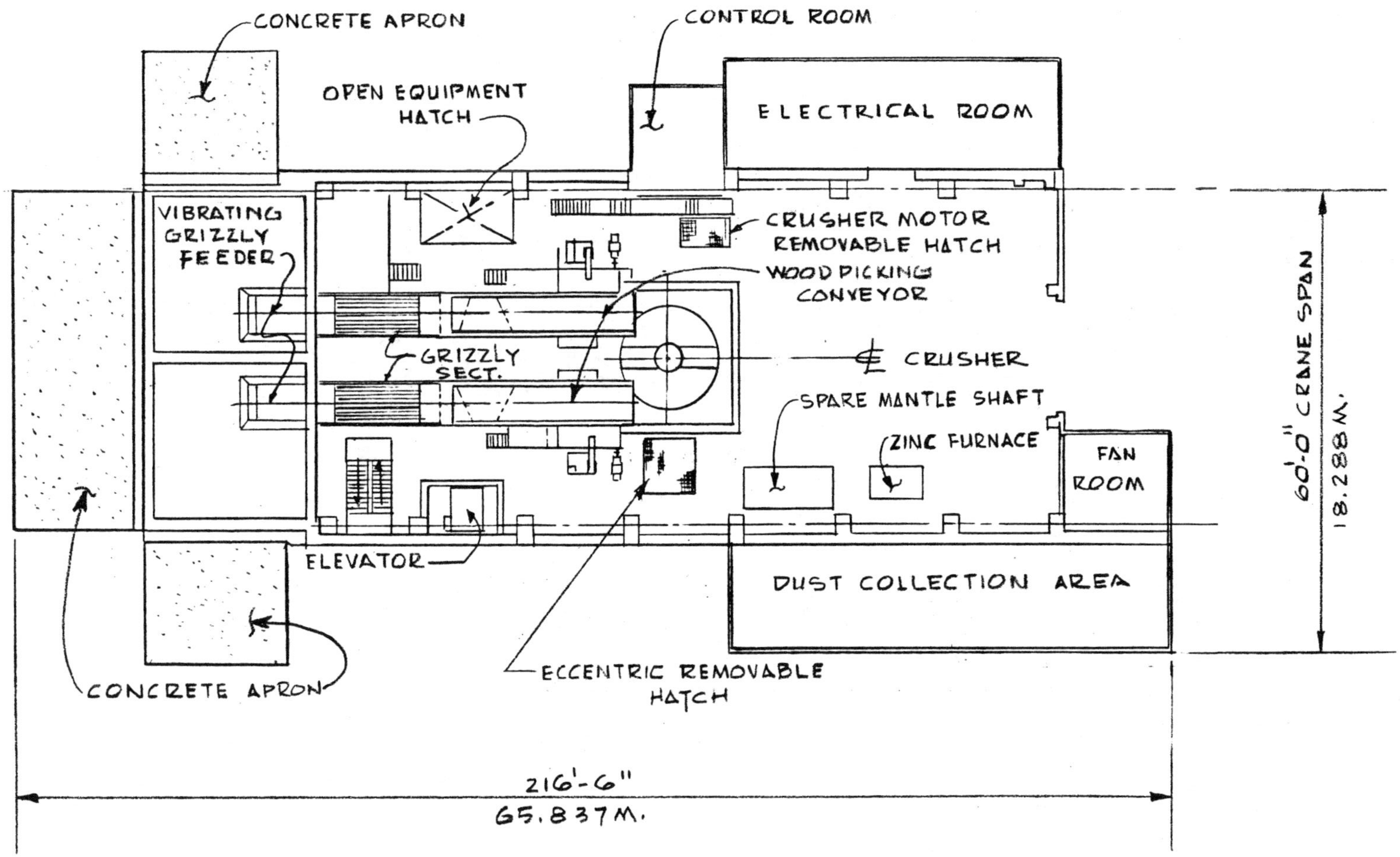

PLAN AT EL. 0'-0" AND EL. +48'-0"

CITIES SERVICE

CITIES SERVICE COMPANY
PINTO VALLEY PROJECT
MIAMI, ARIZONA

The Pinto Valley, Arizona, plant of Cities Service Company is one of the newest copper producers in the United States, having started production in 1975. It is located a few miles west of Miami, Arizona, on the site of the Castle Dome Mine, a World War II operation which was shut down and its mill later moved and renamed the Copper Cities plant. The Pinto Valley concentrator uses six 18 ft (5.5 m) diameter by 21 ft (6.4 m) long, 4000 hp, single-stage ball mills, taking crushed ore from four 7 ft (2.13 m) standard and eight 7 ft shorthead crushers. It has a design capacity of 40,000 stpd but has higher tonnage capability for an expanded concentrator.

The copper porphyry ore is mined by two 15 yd and one 20 yd shovels (11.5 and 15.3 cu m) and is transported to the primary crusher by 150-ton trucks. The dump pocket is relatively shallow (14 ft or 4.27 m deep); and because of the width necessary forthe 150-ton trucks, it was designed to use the ore itself instead of manganese steel as an abrasion-resistant lining, although a heavy steel cap protects the dump pocket top edge. The pocket has a capacity of 300 tons of ore. Water sprays under the dumping positions direct a fog toward the truck being dumped for excellent dust control. Breaking of oversize is done by a hydraulic impact breaker mounted on a hydraulically operated articulated arm situated at the side of the dump pocket. Again because of the shallow pocket, no splitter is used to protect the spider of the 60 in. by 89 in. Traylor gyratory crusher, which is equipped with a hydroset mechanism.

Crushed ore at 8 in. (203 mm) drops into the 360 st surge chamber lined with AR plate and then is fed directly to the 60 in. (1.52 m) stockpile conveyor by an 84 in. by 20 ft (2.13 m by 6.1 m) four-speed apron feeder at a rate of up to 4000 stph. Early in 1977, Cities Service will replace the four-speed electric drive with a hydraulic pump and motor with infinitely variable speed. They are also studying the replacement of the apron feeder with an 8 ft wide (2.44 m) belt feeder similar to those at Similkameen and Palabora. Peak crushing capacity would then be increased to 5000 stph.

The AR plate in the crushed ore pocket was installed in large, rectangular sheets which required much less embedded metal to hold it in place than would the more conventional, smaller and square pieces of cast manganese steel. For duty in the crushed ore pocket, where wear is not severe, the AR plate is a satisfactory material. The new Palabora Figure-8 crusher also uses this design.

Level indication in the crushed ore pocket is accomplished by radiation emitters and sensors with the former being suspended in a pipe in the enclosure wall rather than behind a false wall as has been done in some other installations. Sensors are located on the wall near the eccentric removal door.

The Pinto Valley crusher enclosure is in plan a square and a rectangle, with the rectangle being the maintenance and hoistway area. This two-compartment design permits maintenance to be done easily, and it used only 4870 cu yd (3724 cu m) of concrete in comparison to the three-compartment design at Lornex, which used 6225 cu yd (4760 cu m) of concrete.

Because of the topography near the crusher, the crushed ore conveyor tunnel daylights at an elevation slightly below the lowest level of the crusher enclosure, which permits good drainage. The tunnel itself is made of corrugated and galvanized bolted steel plate. A vertical curve in the conveyor then elevates the ore to a point over the coarse ore stockpile. The conveyor drive consists of two 400 hp motors and one 600 hp motor, with provision for the addition of a second 600 hp motor. Dust at the discharge point of this conveyor is effectively suppressed by a ring of high-pressure water sprays around the falling ore stream just below the head pulley.

Dust collected in the crusher enclosure is drawn downward to a main duct located above the crushed ore conveyor in the conveyor tunnel. The dust collector, which is located near where the tunnel daylights, is of the high-pressure water spray/low air-pressure drop type. Slurried dust is pumped to the concentrator.

CITIES SERVICE COMPANY

Miami, Arizona

Ore type	Copper - Porphyry, 111# / cu ft (1778 kg/m^3)
Concentrator capacity	40,000 st/day (design)
Shovel sizes	Two 15 yd and one 20 yd (11.5 and 15.3 m^3)
Truck size	150 st
Primary crusher	60" x 89" Traylor Gyratory
Connected and consumed horsepower	500 - variable
Capacity of dump pocket	300 st
Grizzly	Not used
Method of breaking or removing oversize	Rock breaker on hydraulic arm
Open side setting (product size)	8 in. (203 mm)
Eccentric throw	1 1/2 in. (38 mm)
Crusher drive	Direct
Capacity of crushed ore pocket	360 st
Crushed ore feeder	84 in. by 20 ft (2.13 m by 6.1 m)
Takeaway belt	60 in. at 530 fpm (1.52 m at 161.5 mpm)
Operating data:	
Shifts operated per week	20
Effective operating hours	6 1/2 per shift
Nominal capacity	3,500 st/hour (design)
Maximum capacity (peak)	4,000 st/hour
Effective availability	95% of scheduled time
Backing for liners	Epoxy
Mantle life	5,500,000 st
Concave life	21,000,000 st
Time to change mantle	8 hours
Time to change concaves	102 hours
Liner materials in dump pocket and crushed ore pocket	None and AR plate
Capacity of overhead service crane	80/20 st
Coarse ore stockpile, after crusher	33,000 st live; 159,000 st total
Direct operating costs per ton (includes labor, power, supplies, maintenance)	$0.038/st (June 1976)
Concrete	4,874 cu yds (3727 m^3)
Reinforcing bar	481 tons
Structural steel	108 tons
Miscellaneous steel	48 tons
Overall height of installation	86' 5 3/4" (26.36 m)
Constructed cost (excludes crushed ore tunnel and conveyor)	ca $5,250,000 (December 1974)

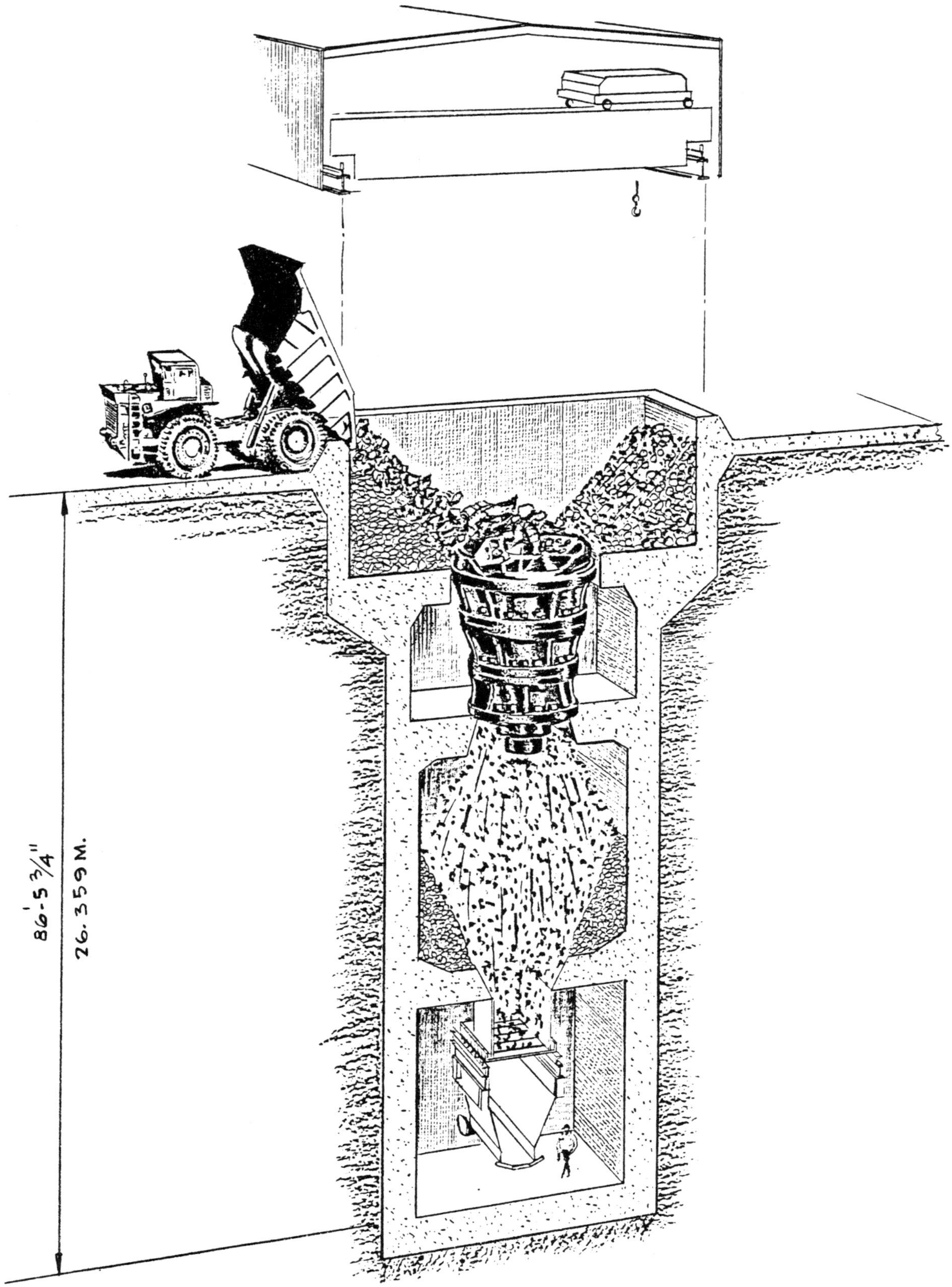

CITIES SERVICE CO
PINTO VALLEY PROJECT
MIAMI, ARIZONA

DUVAL CORP.

DUVAL SIERRITA CORPORATION
TUCSON, ARIZONA

Duval Corporation, the copper-producing subsidiary of Pennzoil United, Incorporated, operates two plants south of Tucson, Arizona: Esperanza and Sierrita. The newest, the Sierrita mine and mill, came into full production in 1971 at a rate of 72,000 tpd (now 87,000 tpd) which at that time made it the largest designed copper plant in the world. In 1968, the largest milling units then available were selected; these included fourteen $16\frac{1}{2}$ ft by 19 ft long ball mills (5 m by 5.9 m), each equipped with a 3,000 hp motor.

The primary crushing plant contains two 60 in. by 89 in. Allis-Chalmers gyratories; each is constructed for two-side dumping by 120- and 150-ton trucks. Access to the maintenance areas to service the drive motors, eccentrics, apron feeders, and speedup conveyor is through hatches between the crushers. This area also contains a good-sized elevator and a stairway.

The rather shallow dump pocket, the 300-ton crushed ore pocket, and the overall design make for a fine, compact arrangement for two 60 in. crushers installed in a total height of 89 ft (27.13 m) from dump level to top-of-concrete at the lowest level beneath the collecting or speedup belt. This efficient layout is reflected in the short changeout time for mantle and concaves (8 and 30 hours respectively) and the low crushing cost of $0.0416 per ton (first half of 1975).

The crushers are installed on 64 ft (19.5 m) centers (Bougainville and Palabora, both with 54 in. crushers, are 55 and 43 ft or 16.76 and 13.1 m respectively). The switchgear room is located between the truck dumping positions on one side of the installation and the hydroset units on the other side, with dust collecting equipment mounted over the latter. Protection of these units from rocks falling from the trucks is provided by concrete walls. Truck dumping is controlled from a central 25 ft high (7.6 m) control room, which permits fine observation of operations. Removal of oversize boulders from the dump pockets is done with a hydraulic grapple or a rock hook or sling attached to the auxiliary crane hook. Spare mantle stands and a maintenance room with a zinc melting furnace are located at one end of the surface facilities. The maintenance room is equipped with an access door for the crane, which is equipped with 80- and 20-ton hoists.

The two 84 in. (2.135 m) apron feeders receiving the 6 in. (152 mm) crushed rock discharge onto a 72 in. (1.83 m) collecting and wear belt operated at a speed compatible with the 60 in. (1.524 m) takeaway belt running at a speed of 630 fpm (192 mpm). This belt feeds a 54 in.

(1.37 m) overland conveyor with a design capacity of 5,000 tph. The overland conveyor transports the crushed ore 12,345 ft (3764 m) to the 40,000 st live, 140,000 st total, coarse ore stockpile adjacent to the fine crushing plant.

In July 1976 construction was started on a similar double crusher installation in the pit. The new crushers will crush ore and the original crushers will be converted to crush waste for transport to the dumps by a new conveyor system. The objectives of the twin in-pit crushers are to reduce the steadily lengthening ore haul to the surface primary crushers, to eliminate 350 vertical feet of truck haul from the bottom of the pit to the crushers, and to eliminate the long truck haul to the waste dump.

The two new 60 in. by 89 in. Allis-Chalmers gyratory crushers will be installed in a building consisting of an underground structure 100 ft high (30.5 m), 50 ft wide (15.25 m), and 100 ft long (30.5 m) and a steel superstructure 50 ft high (15.25 m) and 200 ft long (61 m) with an overhead 80-ton crane, control room, man elevator, etc. Nine thousand cubic yards (6881 m^3) of reinforced concrete and 460 tons of structural and plate steel are being used in the crusher enclosure. Main walls are 4 ft thick (1.2 m) and floor slabs vary from 3 to 6 ft (0.9 to 1.8 m) in thickness.

The new crushed-ore conveyor system will rise from the bottom of the crusher structure on a maximum grade of 28.56%. There will be six transfer points on the modified conveying system stretching over 15,000 ft (4572 m) to the fine crushing plant. With a belt 54 in. (1.65 m) wide, the system will have a rated capacity of 6,450 tph. The two-crusher installation is estimated to cost $14.4 million.

In converting the present primary crushers from ore to waste handling, the size of discharge will be increased from 6 in. (150 mm) to 8.5 in. (216 mm), which will increase the crushing rate from 5,000 to 8,000 tph.

DUVAL SIERRITA CORPORATION

Tucson, Arizona

Ore type	Copper - Porphyry 90#/cu ft (1422 kg/m^3)
Concentrator capacity	87,000 st/day
Shovel size	15 yd (11.5 m^3)
Truck sizes	120 and 150 st
Primary crusher	Two 60" x 89" Allis Chalmers Gyratories
Connected and consumed horsepower	500 - ± 450
Capacity of dump pocket	± 250 st
Grizzly	Not used
Method of breaking or removing oversize	Rock hook, grapple, sling, and crane
Open side setting (product size)	6 in. (152 mm)
Eccentric throw	1 3/8 in. (35 mm)
Crusher drive	Direct
Capacity of crushed ore pocket	300 tons live
Crushed ore feeder	84" apron, 26-49 fpm (2.13 m, 7.9-14.9 mpm)
Speed-up or wear belt	60" - 630 fpm (1.52 m - 192 mpm)
Takeaway belt	60" - 630 fpm (1.52 m - 192 mpm)
Operating data:	
Shifts operated per week	20
Effective operating hours	± 7 per shift
Nominal capacity	2,500 st/hour each
Maximum capacity (peak)	3,200 st/hour each
Effective availability	86% - includes overland conveyor system
Backing for liners	Zinc and epoxy
Mantle life	85 days
Concave life	170 days
Time to change mantle	8 hours - includes pocket clean out
Time to change concaves	30 hours - includes pocket clean out
Liner materials in dump pocket and crushed ore pocket	1 inch AR Plate (25.4 mm)
Capacity of overhead service crane	70 tons - 20 ton auxiliary
Coarse ore stockpile (after crusher)	± 40,000 st - live ± 140,000 st - total
Direct operating costs per ton (includes labor power, supplies, maintenance)	\$0.0416/ton (January - June 1975)
Concrete	8,000 cu yd (6117 m^3)
Reinforcing Bar	800 tons
Embedded Steel	20 tons
Structural steel (prinicpally above truck dump elevation)	380 tons
Overall height of installation	89 ft 0 in. (27.13 m)

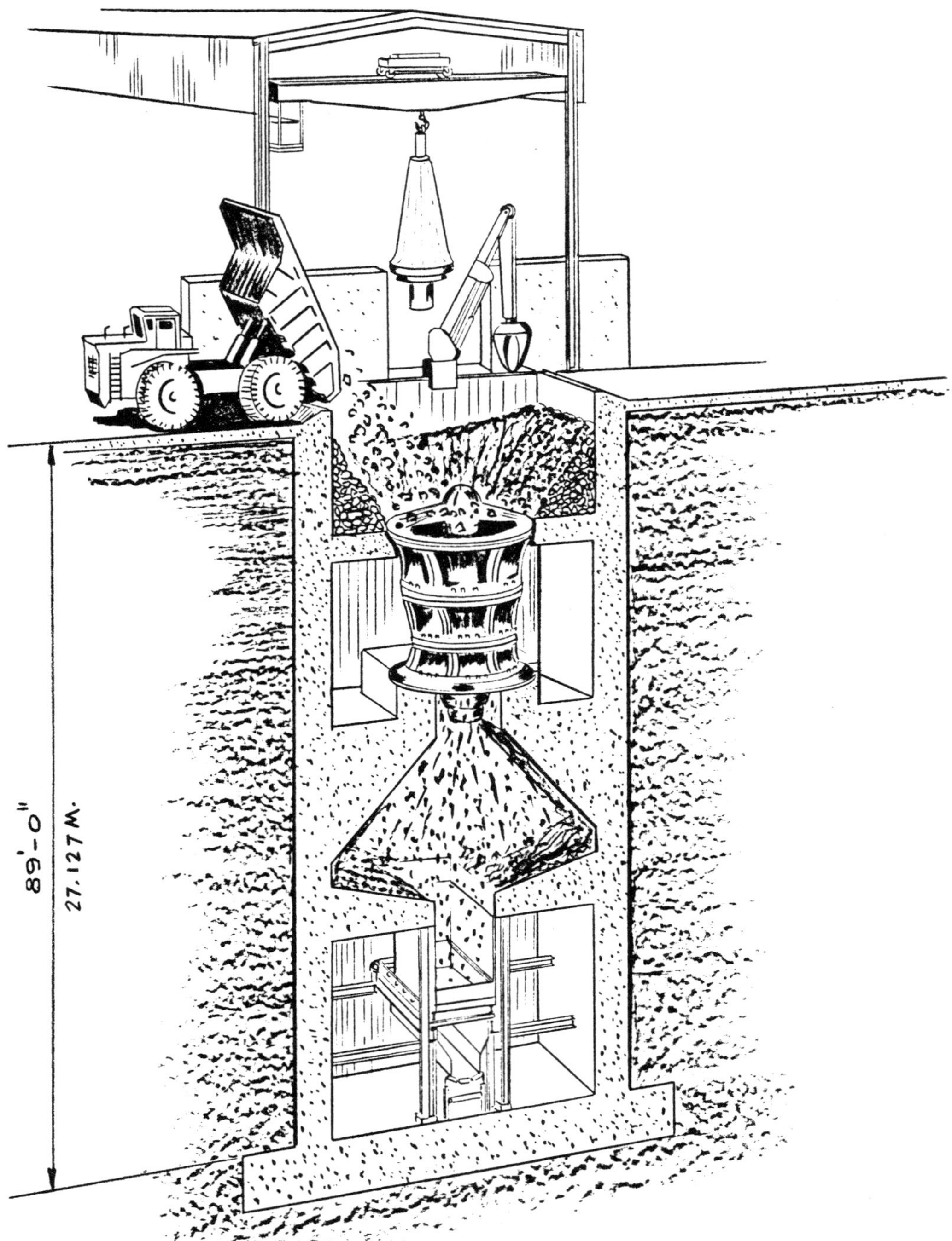

DUVAL SIERRITA CORP.
TUCSON, ARIZONA

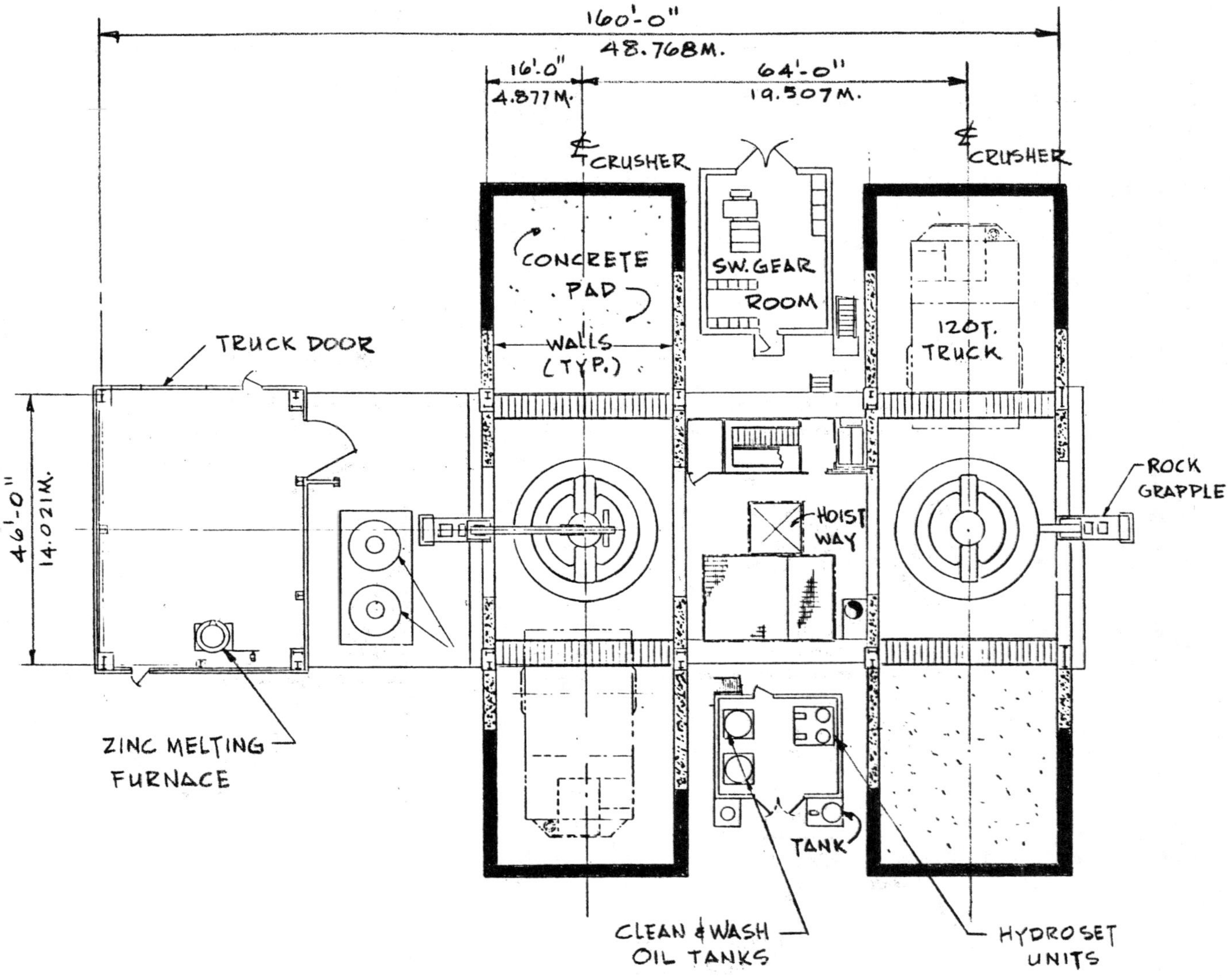

DUVAL SIERRITA CORP.
TUCSON, ARIZONA

IRON ORE CO. OF CANADA

IRON ORE COMPANY OF CANADA
CAROL CONCENTRATOR
LABRADOR CITY, NEWFOUNDLAND

The Carol concentrator of the Iron Ore Company of Canada was completed in 1962 and through expansions has become one of the largest iron ore beneficiation plants in the world with a new feed capacity of 147,000 lt (149,350 mt) per day. The Carol ore contains specular hematite and magnetite having a bulk density of 166 lb/cu ft (2660 kg/m^3). It is crushed to 7 in. (178 mm) for feeding to autogenous mills for subsequent concentration by Humphrey's spirals.

Mine ore is loaded into 150 lt (152 mt) trucks by 10 cu yd (7.65 m^3) shovels and then, at a transfer station, is loaded into side-dump rail cars of 100 lt (101.6 mt) capacity. Each train is composed of twenty cars and operates completely automatically as it hauls the ore to the primary crushing station, dumps its loads, and returns to the loading station, a 11.5 mile (18.5 km) round trip.

The primary crushing station is composed of two 60 in. by 89 in. gyratories installed in 1962 and 1972. Neither crusher has a grizzly ahead of it. Rail cars are dumped into 300 lt (305 mt) pockets which contain splitters for spider protection. What few oversize boulders which do not go through the crushers are drilled and blasted. Liners in the dump pocket and the crushed ore pokcet under the crusher are manganese steel.

Both crushers are direct driven by 700 hp (515 kw) motors, and the hardness of the ore causes them to pull almost their full power. Crushed ore drops into a 160 lt (163 mt) pocket equipped with radioactive level control units for low-intermediate, and high level indications. The low level unit stops and starts the apron feeder, the intermediate unit controls the feeder speed, and a high-level reading stops the car dumping. Apron feeders are 72 in. (1.82 m) wide and are variable speed from 12 to 52 fpm (3.7 to 15.8 mpm). They discharge to 60 in. (1.52 m) wide takeaway belts. Crushing capacity, nominal and maximum for each crusher, is 4,000 ltph (4.064 mtph) and 4,600 ltph (4,674 mtph).

The crusher eccentric is removed by a conventional cart which travels on removable rails. Mainshaft removal and maintenance is done by 65 and 75/10 ton cranes. Mainshaft replacement requires eight hours, a usual time, but concave replacement requires only sixteen hours, a result of a fine maintenance organization. Mantle relining is done at grade next to a repair track. Zinc is used for backing both mantles and concaves.

Both the electrical rooms and the dust collectors are located below grade, with the former being under the rail car dumping point. Dust is drawn upward from the various collection points to the wet collector. Cleaned gasses are sent to stacks by fans located at grade by 23 ft (7 m) below the dump level. Access to all levels of the two crusher installations is by both stairway and elevator. The entire crushing plant is enclosed by a building and doors across the railroad tracks can be closed during maintenance operations. Overall depth of the crusher enclosure is an economical 97 ft 6 in. (32.76 m).

IRON ORE COMPANY OF CANADA
CAROL CONCENTRATOR
Labrador City, Newfoundland

Ore type	Iron ore - specular hematite, 166 lb/ft^3 - 2660 kg/m^3
Concentrator capacity	147,000 ltpd - 149,350 mtpd
Shovel size	10 cu yd (7.65 m^3)
Truck size	150 lt (152 mt)
Rail car size	100 lt (101.6 mt)
Grizzly	None
Capacity of dump pocket	300 lt (305 mt)
Method of removing oversize	Drill and blast
Primary crusher	Two 60 in. by 89 in. Allis Chalmers gyratories
Open side setting	7 in. (178 mm)
Eccentric throw	#1, $1\frac{1}{4}$ in. (31.8 mm); #2, 1-3/8 in. (35 mm)
Crusher drive	Direct - 700 hp installed, 657 hp used
Capacity of crushed ore pocket	160 lt (163 mt)
Crushed ore feeder	72 in. by 14 ft 6 in. (1.83 m by 4.42 m) apron feeder 12 to 52 fpm (3.7 to 15.8 mpm)
Takeaway belt	60 in. (1.52 m); 600 fpm (183 mpm)
Operating data:	
Shifts operated per week	$19\frac{1}{2}$
Effective operating hours	8 per shift
Nominal capacity	4,000 lt - 4,064 mtph per crusher
Maximum capacity (peak)	4,600 lt - 4,674 mtph per crusher
Effective availability	82.5% of scheduled time
Backing for liners	Zinc
Mantle life	#1 - 7 million lt (7,112,000 mt) #2 - 5.6 million lt (5,690,000 mt)
Concave life	#1 - 5.5 million lt (5,588,000 mt) #2 - 5 million lt (5,080,000 mt)
Time to change mantle	8 hours
Time to change concaves	16 hours
Liner materials in dump pocket and crushed ore pocket	Manganese steel
Level controls in crushed ore pocket	3 radioactive units: Low level - feeder start and stop Inter. level - feeder speed High level - dumper stop
Capacity of overhead service crane	75/10 ton and 65 ton
Coarse ore pile after crusher	330,000 lt (335,000 mt)
Date of completion	#1 - 1962; #2 - 1972
Overall height of installation	97 ft 6 in. (32.76 m)

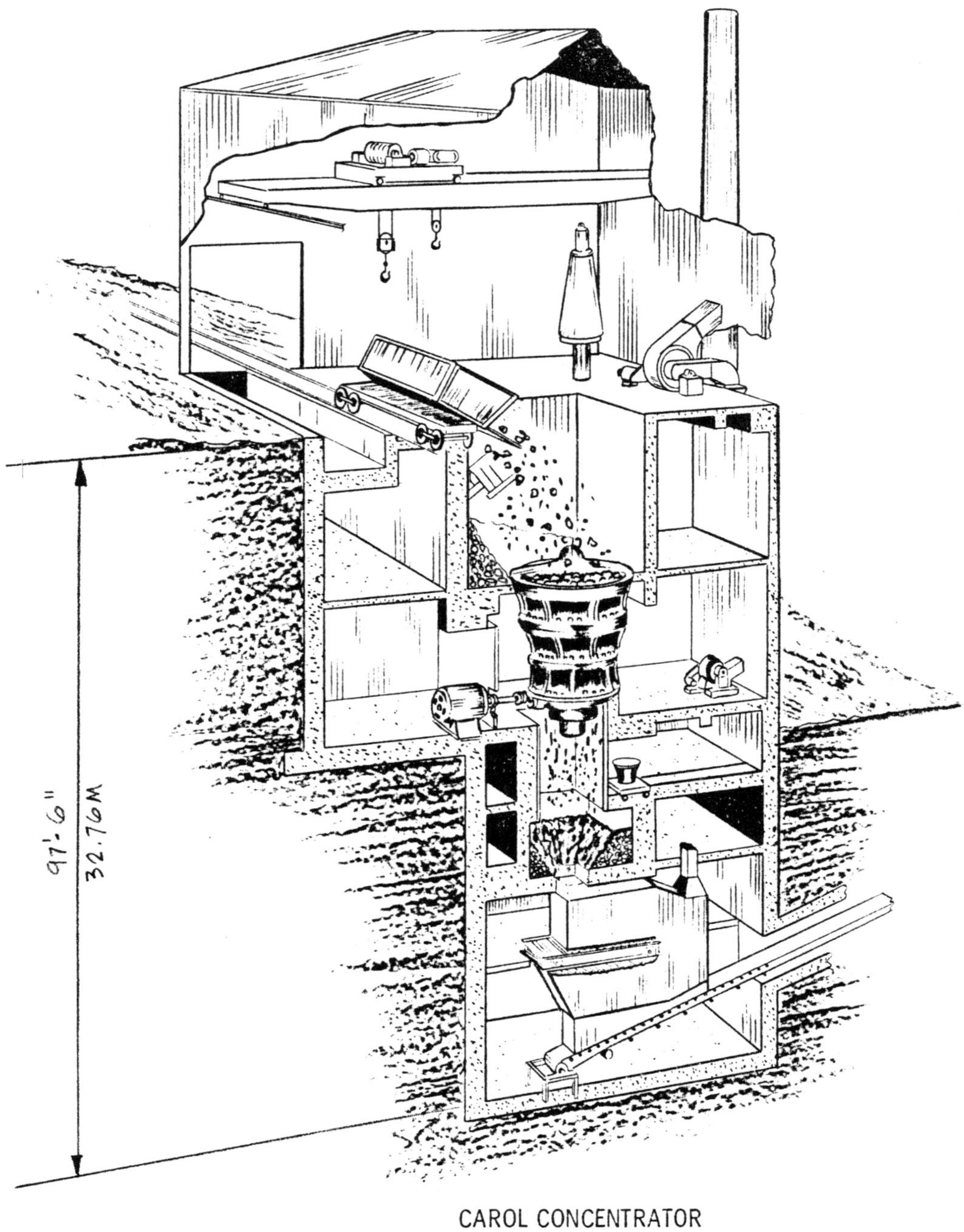

CAROL CONCENTRATOR
LABRADOR CITY, NEWFOUNDLAND
IRON ORE CO. OF CANADA

LORNEX MINING CORPORATION, LIMITED
LOGAN LAKE, BRITISH COLUMBIA

The Lornex mine is located in the Highland Valley area of British Columbia, Canada, one of the most promising districts of low-grade porphyry-type copper mineralization in the world. Lornex is operated by Rio Algom Mines, Limited, a member of the Rio Tinto Zinc Group of London.

Exploration from 1965 to 1968 disclosed an ore body of 293 million tons with a grade of 0.43% copper and 0.14% molybdenum; this led to the decision to construct a 38,000 tpd operation, and production commenced in the fall of 1972. The ore is open-pit mined with 15 cu yd (11.5 m^3) shovels, which load 100- and 200-ton trucks. These trucks dump directly into a 60 in. by 89 in. Allis-Chalmers gyratory from two sides of the 14 ft 6 in. (4.42 m) deep dump pocket and across splitters mounted on the pocket sides. The unlined pocket would hold 1,000 tons level full. The crushed ore discharges into a 300-ton surge hopper equipped with an automatic sonic level indicator to prevent over-filling. A 96 in. by 23 ft (2.44 m by 7 m) apron feeder delivers the ore from the surge hopper to a 72 in. (1.83 m) transfer belt equipped with a tramp iron metal detector.

The transfer belt discharges to a 60 in. (1.52 m) conveyor which conveys the ore about 3,000 ft (914 m) to the 475,000 ton ore storage pile located near the concentrator. Four 48 in. (1.22 m) pan feeders transfer the ore to the mill feed conveyors, which in turn supply the two 32 ft by 15 ft (9.75 m by 4.57 m) semi-autogenous mills, which easily accept the 10 in. (254 mm) product of the crusher. Each semi-autogenous mill is followed by a secondary ball mill. The unusually large open side setting of the crusher accounts for its 6,300 tph maximum capacity. Concentrator capacity is now 48,000 tpd.

Eccentric removal is eased by the design of the door to the crushed ore pocket, which has the cart rails mounted on its outer surface and is lowered to a horizontal position inthe pocket by an electric winch. The top of the door rests on wear bar covered concrete piers in the pocket. Another feature is a chute through the floor at the crusher base that extends into the crushed ore pocket. Dust and spills of ore on that floor during crusher repairs are easily swept into the crushed ore pocket. This feature is also used at Bougainville and Palabora and has been added to others after operation has commenced.

The design follows the concrete rectangular box concept with four main service and maintenance levels shown in the sectional drawings, although a third additional section has been added at the maintenance end. This area accomodates the electrical room and the dust collector at elevation minus 62 ft (18.9 m) which draws dust down through the installation and deposits it on the 72 in. wide (1.83 m) wear belt. Above this area at the surface is an enclosed maintenance area which has a tall door that permits the 80-ton crane to move the mainshaft in for relining. Lornex originally used zinc for relining but have changed to epoxy.

The Lornex primary crusher is an efficient layout utilizing a height of only 85 ft (25.9 m) from the truck dumping roadway to the floor of the transfer belt fed by the apron feeder. The control room is situated so as to allow the operator good vision of the trucks dumping. Oversize boulders are broken with a hydraulic hammer operated from a moveable 25 ft (7.62 m) beam by the control room operator.

LORNEX MINING CORPORATION

Logan Lake, British Columbia

Ore type	Copper-Porphyry, 161#/cu ft in place (2579 kg/m^3)
Concentrator capacity	48,000 st/day
Shovel size	15 yd (11.5 m^3)
Truck sizes	100 and 200 st
Primary crusher	60" x 89" Allis Chalmers Gyratory
Connected and consumed horsepower	700 - 300
Capacity of dump pocket	1,000 st, based on pocket being filled
Grizzly	Not used
Method of breaking or removing oversize	Hydraulic hammer on 25 foot beam
Open side setting (product size)	10 in. (254 mm)
Eccentric throw	1 13/16 in. (46 mm)
Crusher drive	Direct
Capacity of crushed ore pocket	300 st
Crushed ore feeder	96" x 23' apron feeder; variable speed, maximum 51.9 fpm (2.44 by 7 m @ 15.8 mpm)
Takeaway belt	60" at 513 fpm (1.52 m @ 156 mpm)
Wear belt	72" at 345 fpm (1.83 m @ 105 mpm)
Operating data:	
Shifts operated per week	20
Effective operating hours	6.5 per shift
Nominal capacity	4,000 st/hour
Maximum capacity (peak)	6,300 st/hour
Effective availability	95% of scheduled time
Backing for liners	Epoxy
Mantle life	8,500,000 st
Concave life	17,000,000 st
Time to change mantle	10 to 12 hours
Time to change concaves	72 hours
Liner materials	Dump pocket: dead bed Ore pocket: concrete and dead bed Chute liners: NiHard
Capacity of overhead service crane	75 tons
Coarse ore stockpile	475,000 st total
Direct operating costs per ton	Not available
Concrete	6,225 cu yds (4760 m^3)
Structural steel	252 st
Reinforcing bar	670 st
Overall height of installation	85' 2 3/4" (25.98 m)

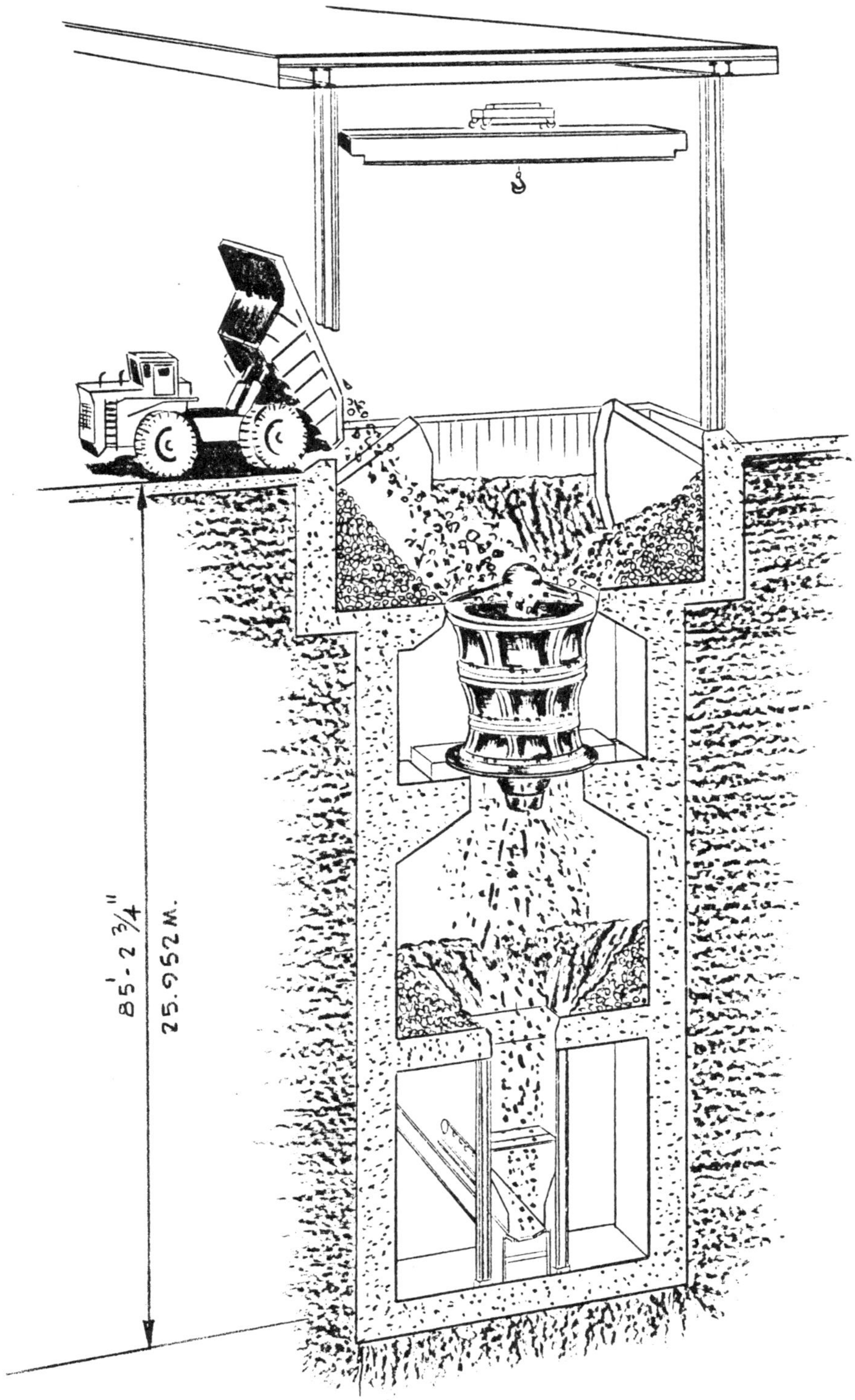

LORNEX MINING CORP. LTD.
LOGAN LAKE, BRITISH COLUMBIA

RESERVE MINING COMPANY
BABBITT, MINNESOTA

The concentration of magnetic taconite ores was first studied by the University of Minnesota in 1913, and the first commercial plant constructed to treat these fine-grained and disseminated magnetites commenced operation in June 1922. One of the partners in this venture, called the Mesabi Iron Company, was Daniel C. Jackling, who pioneered the large-scale treatment of the low-grade copper ores of the West. The Reserve Mining taconite mine at Babbitt, which went into operation in 1955, thirty-three years later, was located on the site of this first taconite concentrator. Although the Mesabi iron plant closed two years later in 1924, it did produce about 400 tons per day of 61 to 64% Fe sinter made from magnetite concentrates.

It is interesting to note the Mesabi Iron crushing plant was similar in design to the later Reserve plant. In that entirely gravity flow plant, an apron feeder fed a 48 in. by 72 in. (1.22 m by 1.83 m) jaw crusher which was followed by a 36 in. by 54 in. (0.91 m by 1.37 m) jaw crusher and then by four 20 in. (0.5 m) gyratories. The structure enclosing these units was 80 ft (24.4 m) deep into the ground.

After years of experimentation by the Mines Experiment Station of the University of Minnesota and many of the mining and iron-and-steel companies, Reserve Mining Company, owned by Armco and Republic Steel, established a 33,000 ltpd mine at Babbitt, Minnesota, and a concentrator and pellet plant at Silver Bay, 47 mi (75 km) by rail on the shore of Lake Superior. The decision to separate the mine and concentrator was based on the need of an ample water supply and an area for tailings disposal. Disposal of the tailings in the lake has been the subject of controversy with environmental groups and authorities, and Reserve is now committed to disposal of these tailings on land.

At the Reserve pit, 11 and 12 cu yd (8.4 and 9.2 m^3) shovels load 85 lt side-dump trucks, which in turn are dumped by a crane directly into a 60 in. by 109 in., 1000 hp Allis-Chalmers gyratory for reduction to $8\frac{1}{2}$ in. (216 mm). The use of these side-dump trucks is now unusual in the mining industry. Oversize rocks are removed by the overhead crane. A 60 in. by 109 in. gyratory at any specific open side setting has the largest capacity of any gyratory crusher. Crushed ore falls into a 150 lt pocket, where it feeds by gravity into four 30 in. (762 mm) secondary gyratory crushers. These units are arranged in a semicircle below and to one side of the primary. The feed pocket of each crusher is a separate compartment to permit maintenance operations by the overhead crane. Secondary crushed ore is combined in another pocket where it is withdrawn by an 84 in. (21.3 m) apron feeder for transfer to the 60 in. (1.52 m) takeaway conveyor operating at 380 fpm (116 mpm). This takeaway conveyor, in turn, delivers the ore to loading bins for train transport to the concentrator.

Access to the four secondary gyratories and other equipment is through hatches in the several floors of the structure. The 130-ton crane with a lighter capacity fast hook provides lifting capability for all necessary maintenance functions. Even with the extremely heavy mantle of the primary (105 tons), mantle changeout time is only 12 to 16 hours. Well-organized maintenance procedures also permit concave replacement to be done in only 32 to 44 hours. Even with the extremely abrasive taconite which limits concave and mantle life to 3,500,000 tons, operating availability is a remarkable 93%. Peak crushing capacity is 5000 ltph.

In design, the crusher enclosure is circular on the upper levels and more than half a circle on the lower levels; the inside diameter is 74 ft (22.6 m). The efficiency of using concrete in this way (see the descriptions of the circular Inspiration and Figure-8 Palabora crushers) is demonstrated by the 11,000 cu yd (8411 m^3) of concrete used in this installation, which is 165 ft (50.3 m) deep, about twice the normal height of 60 in. (1.52 m) crusher installations. Included in this yardage are the takeaway conveyor tunnel, conveyor drive house and gallery, and the loading bins. In contrast, Anaconda's Butte crusher enclosure, which is of rectangular design and is 122 ft (36.57 m) deep, used 12,800 cu yd (9787 cu m) of concrete. Reinforcing bar was also below average at 600 tons or 109#/cu yd (64.8 kg/m^3).

The 1976 crushing cost at Reserve of $0.15 - $0.20 per long ton includes labor, power, supplies, and maintenance for the one primary and four secondaries and attests to very efficient operation on these hard taconite ores.

RESERVE MINING COMPANY

Silver Bay, Minnesota

Ore type	Taconite (iron ore), 130#/ cu ft (2083 kg/ m^3)
Concentrator capacity	90,000 lt/day
Shovel sizes	11 - 12 yd (8.4 - 9.2 m^3)
Truck size	85 lt side dump
Primary crusher	60" x 109" Allis Chalmers Gyratory
Connected and consumed horsepower	1,000 - 700
Capacity of dump pocket	150 - 200 lt
Grizzly	Not used
Method of breaking or removing oversize	Removal by overhead crane
Open side setting (product size)	8 1/2 in. (216 mm)
Eccentric throw	1 1/2 in. (38.1 mm)
Crusher drive	Direct
Capacity of crushed ore pocket	150 lt
Crushed ore feeder	None, gravity to secondary crushers
Takeaway belt	60 " at 380 fpm (1.52 m @ 116 mpm)
Operating data:	
Shifts operated per week	17 1/2
Effective operating hours	7 hours per shift
Nominal capacity	3,000 lt/hour (design)
Maximum capacity (peak)	5,000 lt/hour
Effective availability	93% of scheduled time
Backing for liners	Zinc
Mantle life	3,500,000 lt
Concave life	3,500,000 lt
Time to change mantle	12 to 16 hours
Time to change concaves	32 to 44 hours
Liner materials in dump pocket and crushed ore pocket	Manganese and martensitic cast steel
Capacity of overhead service crane	130 tons
Coarse ore stockpile after crusher	500,000 lt (after loading bins)
Direct operating costs per ton (includes labor, power, supplies, maintenance)	$0.15 to $0.20/ lt (includes secondary crushing and loadout)
Concrete	11,000 cu yd** (8411 m^3)
Reinforcing bar	600 tons **
Structural steel	1,000 tons**
Overall height of installation	165' 0" (50.3 m)

** Includes primary and secondary crusher enclosure, conveyor tunnel, drive house, gallery, and loading bins.

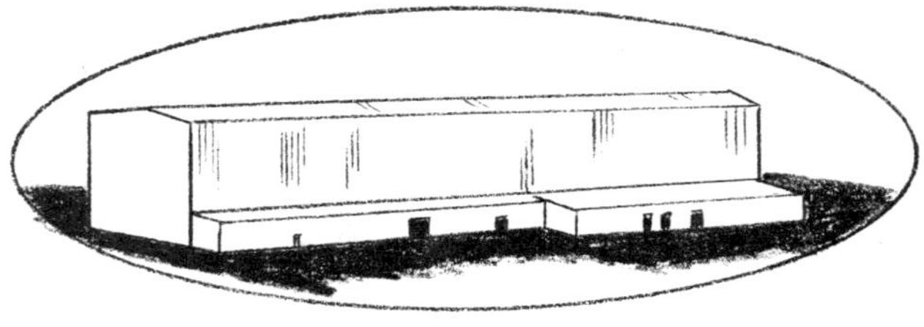

PLANT PERSPECTIVE VIEW

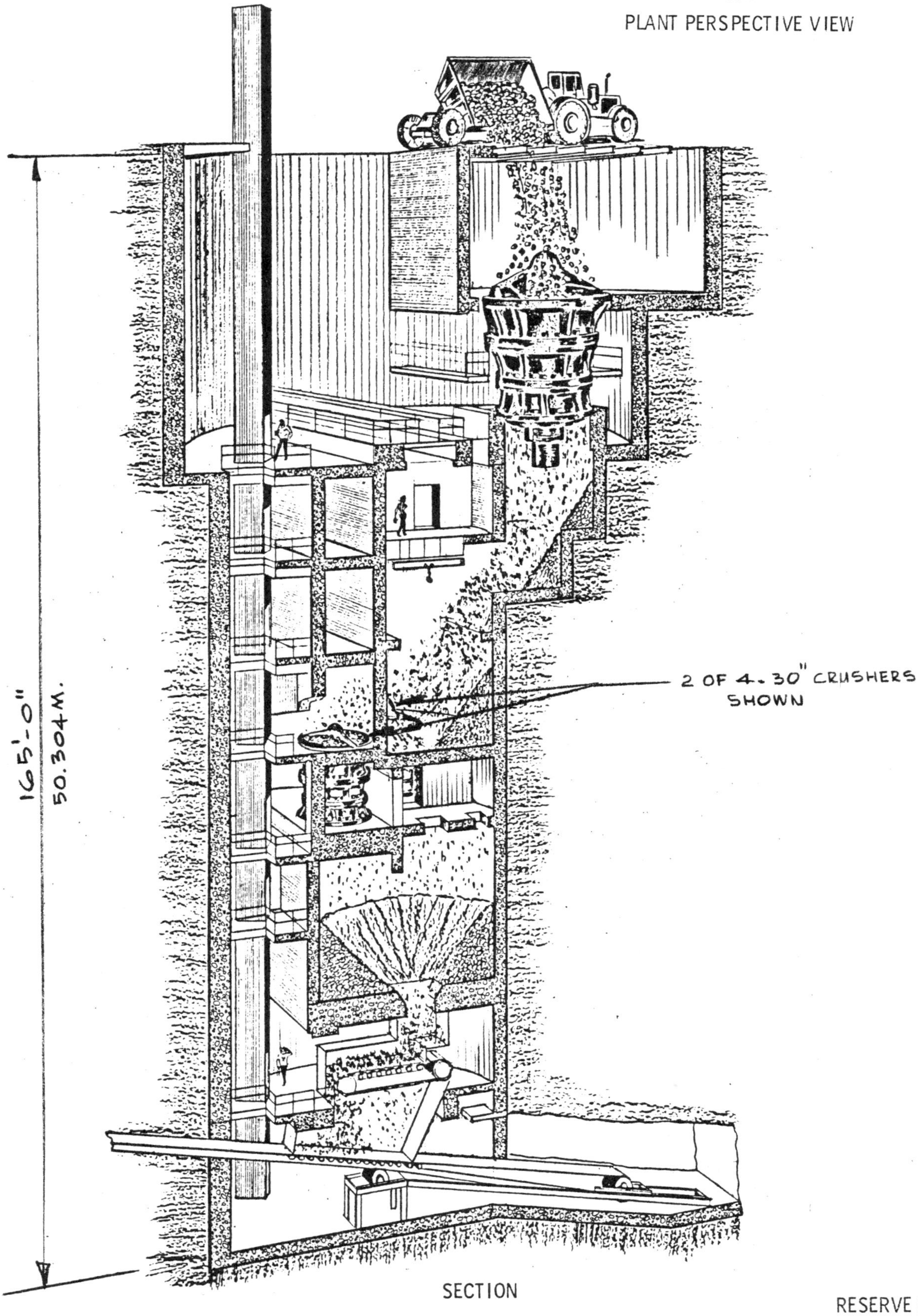

SECTION

RESERVE MINING C
BABBITT, MINN.

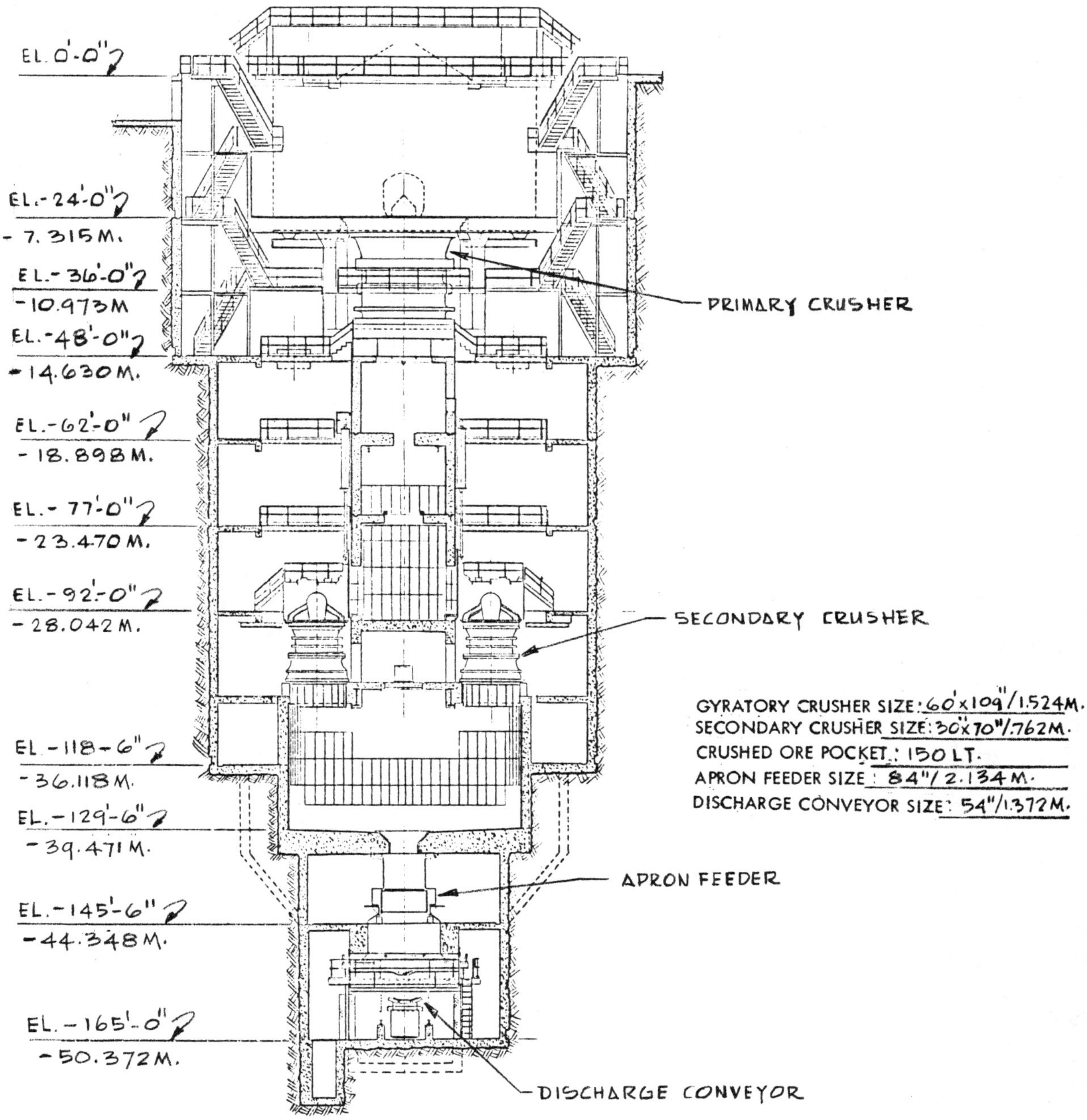

SECTION

RESERVE MINING CO.
BABBITT, MINN.

SOUTHERN PERU COPPER COMPANY
CUAJONE PROJECT, PERU

The Cuajone and Toquepala copper projects are located in the extreme southern area of Peru. Since 1960, Southern Peru Copper Company has operated the Toquepala open-pit mine, flotation mill, Ilo smelter, townsites, power plants, and railroad. Toquepala and Ilo, the smelter and port location, are connected by a 110-mile-long (176 km) railroad. Much of Toquepala's infrastructure has been used for the Cuajone project. Cuajone, at an elevation of 11,400 ft (3475 m), required five standard-guage railroad tunnels with a combined length of 17 miles (27.2 km) to connect with Toquepala so concentrates could be shipped to the Ilo smelter. The total cost of Cuajone is estimated at $726 million for a designed annual capacity of 180,000 tons of copper, equivalent to about $4,000 per ton of installed production capacity. This relatively low cost per ton, by today's standards, has been made possible through the use of the Ilo-to Toquepala railroad, the ability to expand the Ilo copper smelter, and other favorable factors.

To mine the 468 million tons of copper sulfide ore at a grade of 1.0% copper, approximately 241 million tons of waste were stripped before mining of ore to supply the new Botiflaca 45,000 ton per day concentrator. The mining rate must exceed 300,000 tons of waste plus ore per day to maintain the 45,000 tpd sulfide ore milling rate which was to be reached at the end of 1976.

Waste stripping is done with 15 cu yd electric shovels and 120-ton trucks. Ore mining is done with 8 and 9 yd shovels. A railroad connects the pit with the 60 in. by 89 in. Allis-Chalmers gyratory crusher. Side-dump railroad cars of 50 cu yd capacity dump onto a 16 ft 9 in. long (5.1 m) grizzly inclined at 40° with 7 to 8 in. (178 to 203 mm) bar spacings. The grizzly is divided into two 22 ft 4 in. (6.8 m) wide sections and two side-dump railroad cars discharge simultaneously into the 150-ton capacity dump pocket. A Barco rock manipulator is used for breaking or removing oversize from the dump pocket. The bin receiving the grizzly undersize has a capacity of 5,000 tons, and the pocket beneath the crusher holds 900 tons. Each is equipped with an 84 in. apron feeder, which feeds a 60 in. (1.52 m) conveyor belt to the 150,000 ton coarse ore stockpile, where the ore is distributed by a 72 in. (1.83 m) tripper conveyor. The 45,000 live tons in the coarse ore stockpile are reclaimed by twelve 48 in. (1.22 m) apron feeders.

Maintenance on the crusher eccentric is made easy by an eccentric removal cart mounted on rails which are mounted in the upper part of the crushed ore pocket at a guage wider than the fan of ore falling from the crusher. The eccentric is carried below the cart wheels rather than above. A similar, permanently mounted eccentric cart is used at Climax's new Henderson crusher. Maintenance is also aided by the 80-ton bridge crane with an auxiliary 20-ton fast hook. The wet-type dust collector is mounted a few feet below the railroad dump level, and dust is drawn upward to this unit.

It is interesting to compare the 124 ft 10 in. (38.06 m) height of the Cuajone crusher with others. The 54 in. by 74 in. railroad dump crusher at Bonneville is 139 ft (42.37 m), which is deeper than the Cuajone crusher; but it has an additional conveyor and transfer point at the lowest level. Otherwise Cuajone and Bonneville would be within about 2 ft (0.6 m) of each other. The only truck dump gyratory with a comparable enclosure depth is at Anaconda's Butte operation (120 ft or 36.58 m), but the vibrating grizzly feeders and picking belts ahead of that crusher require substantial head room. Other 60 in. truck dump crushers without grizzlies which can be compared to Cuajone are Duval-Sierrita at 89 ft (27.12 m), Lornex at 85 ft (25.91 m), and Pinto Valley at 86 ft (26.21 m). The wide dumping pocket at Cuajone, with its twin grizzlies, requires 34 ft (10.36 m) of height. If a Sierrita or Pinto Valley dump pocket had been used, the Cuajone crusher enclosure would have decreased its height to only 104 ft (31.7 m).

The capital cost of the Cuajone crusher in 1976, which involved the placement of 8,400 cu yd of concrete and 560 tons of structural and platform steel, is estimated at $4,531,000 in direct costs with a total cost estimated at $6.4 million, not including freight and duties on equipment and supply items.

SOUTHERN PERU COPPER CORPORATION

Cuajone Project

Ore type	Copper porphyry, 100-110# cu ft (1610 - 1762 kg/m^3)
Concentrator capacity	45,000 st/day
Shovel sizes	8 and 9 yds (6.1 and 6.9m^3)
Railroad car size	50 yd side dump
Primary crusher	60" x 89" Allis Chalmers Gyratory
Connected and consumed horsepower	500 hp - not available
Grizzly	Single divided into two halves
Length and width	16' 9" -- two sections of 22' 4" each (5.1 - 6.8 m)
Slope	40°
Spacing of bars	7 to 8 in. (178 to 203 mm)
Method of breaking or removing oversize	Barco rock manipulator
Open side setting (product size)	8 in. (203 mm)
Eccentric throw	1 13/16 (46 mm)
Crusher drive	Direct
Capacity of dump pocket	Approximately two cars
Capacity of crushed ore pocket	Approximately 5,000 tons under grizzly Approximately 900 tons under crusher
Crushed ore feeders	Two 84" apron feeders (one for crusher and one for grizzly) (2.13 m)
Takeaway belt	60 inch at 650 fpm (1.52 m @ 198 mpm)
Operating data:	
Shifts operated per week	17 planned
Effective operating hours	6.5 hours per shift
Nominal capacity	4,000 st/hour
Maximum capacity (peak)	5,000 st/hour
Effective availability	75% (estimated) of scheduled time
Backing for liners	Zinc
Mantle life	18 months (estimated)
Concave life	24 months (estimated)
Time to change mantle	10 hours (estimated)
Time to change concaves	6 shifts (estimated)
Liner materials in dump pocket and crushed ore pocket	1, 1 1/2, and 2 inch AR plate depending on zone (25.4 - 38.1 - 50.8 mm)
Capacity of overhead service crane	80/20 ton
Direct operating costs (includes labor, power, supplies, maintenance)	Not available
Coarse ore stockpile	45,000 tons live; 150,000 tons total
Concrete	8,400 cu yds (6423 m^3)
Reinforcing bar	1,000 st
Structural and platform steel	560 st
Capital cost	$6,400,000 (not including freight, duties, and imported supply items)
Overall height of installation	124' 10" (38.05 m)

SOUTHERN PERU COPPER CORPORATION

CUAJONE PROJECT

72" GYRATORY

JOHNS-MANVILLE

CANADIAN JOHNS-MANVILLE CO., LTD.
ASBESTOS, QUEBEC

The Jeffrey operation of Canadian Johns-Manville Co., Ltd., at Asbestos, Quebec, is considered to be the world's largest asbestos mine. It accounts for approximately 37% of Canadian production of chysotile asbestos and about 13% of world production. Rated capacity is over 600,000 stpy (540,000 mtpy) of fiber, of which well over 90% is exported to the United States and seventy other countries.

During the early 1970's the company completed a five-year modernization and expansion program which included the installation of a unique 72 in. by 93 in. gyratory crusher to handle the extremely blocky serpentinized peridotite ore. Because of the large diameter of this crusher the top ring of the shell was made in two halves which were bolted together. Hundred-ton trucks dump from two sides of the crusher into the 200 st (180 mt) dump pocket which is equipped with dust collecting ducts in the walls of the other two sides. The installation was designed to accommodate the eventual use of 200-ton trucks. Crushed ore at a size of $7\frac{1}{2}$ in. (191 mm) falls into a uniquely designed crushed ore pocket equipped with three sloping steel sides and one vertical side, which make it totally live. Infrared heaters installed in back of the hopper sides prevent the ore from sticking to the steel. Another interesting feature is an eccentric removal bridge which is permanently mounted inside the crushed ore pocket. The two halves of the bridge are let down by electric winches to form a platform on which the eccentric cart travels and from where maintenance is done. Level indications in the crushed ore pocket are by radioactive indicators.

Under the crushed ore pocket is a 72 in. by 28 ft (1.83 m by 8.5 m) apron feeder which operates at 7.2 to 50 fpm (2.3 to 15.3 mpm). Crushed ore is discharged to a picking belt for tramp removal. A crossbelt magnetic separator is also mounted over this belt. Permanently mounted belt reels are also situated over the picking belt to facilitate the changing of this unit. A 60 in. (1.52 m) wide takeaway belt then elevates the crushed ore to the surface facilities. Instead of discharging to a crushed ore stockpile, the takeaway conveyor delivers the ore to a 300-ton distribution bin. The ore is then screened and secondary crushed; and the plus 3/4 in. (19 mm) is stored in a 3,000 ton bin while the minus 3/4 in. is dried and then sent to storage. This system of directly screening primary crushed ore is similar to that used at both the Endako molybdenum and Gibraltar copper operations in British Columbia. While it would seem connecting processing equipment directly to a primary crusher would have an adverse effect on the crusher availability, the 94% of scheduled time this crusher operates would seem to disprove this theory. It should be noted, however, that the crusher operates 17 shifts per week.

On the floor immediately above the apron feeder are located the crusher lubrication system and the hydraulic unit for the crusher mainshaft. The next floor above accommodates the eccentric cart tracks, and the eccentric is removed through double doors in the upper part of the crushed ore pocket. Both doors are heavily armored for wear resistance. At each level trolley beams are conveniently located for equipment maintenance. The crusher itself is mounted on a concrete diaphragm which spans the distance between the enclosure walls. A stairway as well as an elevator give access to all operating levels.

Above the truck dumping level the crusher enclosure is protected by a 60 ft (18.2 m) high building which houses the 100/20 ton crane, the control room, the zinc furnaces, and the maintenance area. The control room is above and directly to the side of the crusher where it provides an excellent view of the dumping and crushing operations. During maintenance, four-leaf vertical sliding doors can be lowered at each dumping position to contain heat in the building. Vertical sliding doors and a hinged, vertical swinging door at the crane level can be opened so the crane carrying the mainshaft can pass through into the heated maintenance and relining area. Dry dust collectors are located to one side of the maintenance area and over the electrical room, office, lavatory, and lunch room. The roof of the building is typical of those built in heavy snow country in that it drains to the center rather than to the sides. Overall height of the crusher enclosure from dumping level to top of concrete at the bottom floor is 109 ft 6 in. (33.4 m).

Capital cost of this unique crushing plant which was completed in November of 1972 was $4,750,000 (Canadian). Operating cost in the fall of 1977 was $0.03 per short ton.

CANADIAN JOHNS-MANVILLE CO., LTD.

Asbestos, Quebec

Ore type	Asbestos (serpentinised peridotite) 177 #/cu ft (2830 kg/m^3)
Concentrator capacity	45,000 stpd (40,700 mtpd)
Shovel sizes	8 and 10 cu yd (6.2 and 7.6 m^3)
Truck size	100 st (91 mt) can accept 200 st trucks
Primary crusher	72 in. by 93 in. Traylor gyratory
Connected and consumed horsepower	700-600 hp (520-450 kw)
Capacity of dump pocket	200 st (180 mt)
Grizzly	Not used
Method of breaking or removing oversize	By crane using cables attached to rock bolts in oversize rock
Open side setting (product size)	7½ in. (191 mm)
Eccentric throw	1 3/4 in. (44 mm)
Crusher drive	Direct
Capacity of crushed ore pocket	330 st (300 mt)
Crushed ore feeder	72 in. (1.83 m) apron, 7½ to 50 fpm (2.3 to 15.3 mpm)
Picking belt	84 in. by 122 ft (2.13 m by 37.2 m)
Takeaway belt	60 in. wide (1.52 m)
Operating data:	
Shifts operated per week	17
Effective operating hours	7 per shift
Nominal capacity	2300 stph (2080 mtph)
Maximum capacity (peak)	3000 stph) 2720 mtph)
Effective availability	96% of scheduled time
Backing for liners	Zinc
Mantle life	30 million st (27.2 million mt)
Concave life	Not changed yet
Time to change mantle	During plant shutdown (1 week)
Time to change concaves	Not known
Liner materials in dump pocket and crushed ore pocket	Manganese plate
Capacity of overhead service crane	100 and 20 st (91 and 18 mt)
Coarse ore stockpile after crusher	300 st (272 mt) distribution bin to screens
Direct operating costs per ton (includes labor, power, supplies, and maintenance)	\$0.03/st (Canadian funds) (October 1977)
Capital cost	\$4,750,000 (Canadian) (November 1972)
Overall height of installation	106 ft 6 in. (32.4 m)

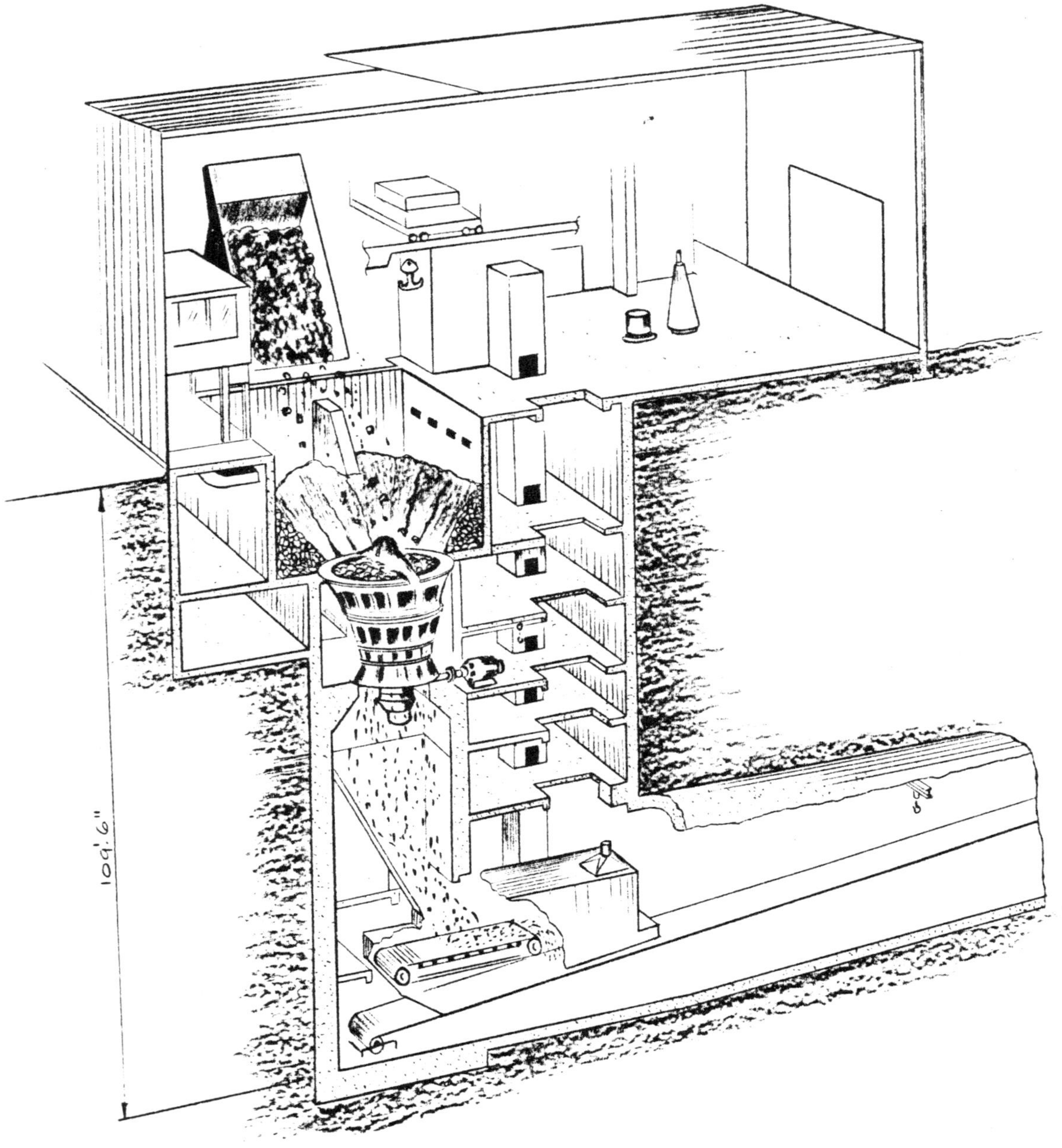

CANADIAN JOHNS MANVILLE CO. LTD.
ASBESTOS, QUEBEC

ASARCO SACATON

ASARCO
SACATON UNIT
CASA GRANDE, ARIZONA

Asarco's Sacaton unit, located near Casa Grande 45 miles southeast of Phoenix, Arizona, was brought into production in February 1974. Ore reserves are estimated at about 48 million tons at 0.95% copper. Mineralization of the orebody comprises both sulfides and oxides, with minor amounts of gold and silver. Preproduction stripping of 40 million tons of overburden was accomplished in less than two years. Mining is done at the rate of 9,000 stpd ore with a waste-to-ore ratio of 4.6 to 1.0. The mine is developed on 40 ft benches, and by 1979 underground block caving methods will be phased into the operation at a depth of 1900 ft.

Ore is mined in the pit two shifts per day, seven days per week, and is hauled with 85 ton trucks to the primary crushing plant. Trucks dump their loads into a 165 st capacity pocket equipped with a chain curtain which controls the ore flow to the 6 ft 6 in. wide by 24 ft long (2.0 m by 7.3 m) vibrating feeder grizzly. Grizzly bars are inclined at a 5° slope on the first half of the feeder and 10° on the last half. Grizzly bars have 6 in. (13 mm) spacings which pass 60 to 70% of the mine-run ore. Grizzly oversize drops into a 48 in. by 60 in. (1.22 m by 1.52 m) primary jaw crusher. Both the grizzly undersize and crusher discharge are collected on a 42 in. (1.07 m) conveyor belt operating 398 fpm (121 mpm) for transport to the 4,750 live st intermediate storage area ahead of the fine crushing plant.

Capacity of the Sacaton primary crushing plant is a nominal 1150 stph (1043 mtph). It operates fourteen shifts per week with an effective crushing time of four hours per shift.

An economical open-air design was used, but an enclosed operator control room is located on the upper level where the dump truck, feeder, and crusher can be observed. Oversize rocks are removed from the feeder or crusher by the 10-ton bridge crane.

Direct operating costs for this well-designed primary crushing plant, including labor, power, supplies, and maintenance, are only 3.5 cents per short ton.

An interesting feature of the Sacaton orebody is that the non-sulfide copper minerals, which are less prevalent than the sulfides, occur intimately with the sulfide minerals; for this reason, all ores as mined are treated in a flotation circuit using sulfidizing methods for oxide recovery.

ASARCO

Sacaton Unit

Casa Grande, Arizona

Ore type	Copper porphyry, 100#/ft^3 (1600 kg/m^3)
Concentrator capacity	9,000 stpd design - 14,000 stpd maximum
Shovel sizes	9 yd (6.9 m^3) shovels, 12 yd (9.8 m^3) front-end loaders
Truck size	85 st (77.1 mt)
Capacity of dump pocket	100 st maximum (91 mt)
Method of breaking or removing oversize	Hooking to reposition or removing with cable choker
Crusher feeder	6 ft 6 in. x 24 ft (1.98 x 7.32 m) vibrating grizzly
Slope	1st 10 ft (3.05 m) 5° - 2nd 14 ft (2.47 m) 10°
Spacing of bars	6 in. (12.7 mm)
Estimated % of ore through grizzly	60-70%
Primary crusher	48 in. x 60 in. (1.22 x 1.52 m) Allis Chalmers jaw
Connected and consumed horsepower	200 - 115 hp (150 - 86 kw)
Open side setting (product size)	7 in. (178 mm)
Eccentric throw	1.75 in. (44.5 mm)
Crusher drive	Belt drive
Capacity of crushed ore pocket	None - drop chute and belt-loading vee
Takeaway belt	42 in. (1 07 m) - 398 fpm (121 mpm)
Operating data:	
Shifts operated per week	16
Effective operating hours	4 hours per shift
Nominal capacity	1150 stph (1043 mtph)
Maximum capacity (peak)	1300 stph (1179 mtph)
Effective availability	94% of scheduled time
Backing for liners	None
Jaw plate life	4 million st (3.63 million mt)
Cheek plate life	20 million st (est) (18 million mt)
Time to change mantle	N/A
Time to change concaves	N/A
Liner materials in dump pocket	Cast manganese steel
Capacity of overhead service crane	10 st (9.1 mt)
Coarse ore stockpile after crusher	8,000 st live, 32,000 st total
Direct operating costs (includes labor, power, supplies, maintenance)	\$0.034 per short ton - June 1977

CARLIN GOLD MINING COMPANY
CARLIN, NEVADA

The Carlin gold deposit is located in northeastern Nevada about thirty-five miles by road from the town of Elko. Following the staking of six square miles of ground, a drilling program outlined a fine ore body mineable by open pitting. The resulting mine turned out to be the largest open-pit gold operation in the world, until the Rosario Pueblo Viejo operation was started in 1975. Because of the high profitability forecast at the then $35 gold price, construction of a 2000 stpd mill was commenced on a crash program and was completed for Newmont Mining Corporation by Bechtel Corporation in ten months. Production was started in early 1965.

The gold mineralization occurs in marine-type sediments, chiefly siltstones and limey shales, and the ore characteristics make for a difficult materials-handling problem. At 5% to 6% moisture the ore is dusty during crushing, but even a light rain or snowfall virtually turns the material into a slurry. Crushing this sticky ore below 3/4 in. (19 mm) in the dry season and 1 in. (25.4 m) during the winter is extremely difficult and leads to problems with conveyors, chutes, and fine ore storage.

The initial flowsheet for Carlin's crushing plant used a 5 ft by 22 ft (1.52 m by 6.7 m) vibrating feeder-grizzly under the 60-ton dump pocket for feeding the 42 in. by 48 in. (1067 mm by 1220 mm) single toggle jaw crusher and a $5\frac{1}{2}$ ft (1.52 m) standard cone crusher in closed circuit with a 6 ft by 12 ft (1.83 m by 3.66 m) vibrating double-deck screen. Because continual packing of fines in the vibrating feeder pan would cause frequent plant shut-downs to remove the fines with jack hammers, the decision was made to replace the vibrating unit with an apron feeder and stationary grizzly ahead of the jaw crusher. A dribble belt was also installed under the apron feeder. This installation has greatly improved the feeding of the pit run ore to the primary jaw crusher.

Mining at Carlin is done by a 5 yd shovel and $7\frac{1}{2}$ and 10 yd front-end loaders, and ore transport is by several sizes of trucks from 30 to 75 tons capacity. These pit haulage trucks dump onto stockpiles adjacent to the primary crusher. Ore is then fed to the dump hopper by front-end loaders only when the crushing plant is operating. This extra handling procedure is necessary because plant delays caused by problems with wet ore in the secondary crusher resulted in the ore setting up or packing in the dump pocket. Solidified material in the dump pocket would in turn delay truck haulage. The substitution of an apron feeder for the vibrating feeder and the use of rubber-covered screens have contributed greatly to overcoming wet ore problems. In addition, blending different types of ores fed to the plant has aided crushing.

A control room is located above the dump pocket where the operator can clearly observe and control the front-end loader feeding rate, apron feeder, jaw crusher, three conveyors, screen, and secondary cone crusher. Because of the sticky nature of the Carlin ores, crushing and screening are important and problem procedures. Through good visual observation and with all equipment controls at hand in the control room, one operator handles all primary and secondary crushing plant functions. Solid-state instrumentation is used for all measurements and control functions brought to the control room panels. The control room is heated by a unit separated from the crushing plant building and is cooled with an Alpine evaporator-cooler mounted on the roof.

The peculiar nature of the wet, sticky Carlin ore which is similar to laterite-type material raises the question of using secondary reduction crushing with the auxiliary conveyors, screens, chutes, etc., following primary crushing. With eleven years of experience in the materials-handling problems of Carlin ores coupled with operating information on similar materials, the authors believe a more suitable flowsheet would encompass semi-autogenous grinding. This flowsheet would eliminate crushing after the primary jaw, with the attendant conveyors, screen, chutes, and more particularly the fine ore bins with their feeders and conveyors and finally the conventional grinding mills. In place of all this equipment would be the apron feeder, stationary grizzly, jaw crusher, and a conveyor which would directly feed a semi-autogenous mill. The ground product at about 48 mesh could be stored in the large-diameter holding tanks equipped with rakes and circulating pumps which are successfully used for this purpose in the cement industry. The slurry would be withdrawn at a measured flow rate for processing or final grinding if necessary. In this type of flowsheet for wet, sticky ores which present problems in a jaw crusher, a Wobbler feeder between the apron feeder and the crusher to remove the fines has proved at times to be a good choice.

Rosario Resources did install semi-autogenous grinding at their 8000 stpd gold and silver plant, Pueblo Viejo, in the Dominican Republic. Pit-run ore is dumped over a grizzly with 16 in. (406 mm) square openings into a 750-ton bin from which it is withdrawn by two 54 in. by 34 ft (1.37 m by 10.36 m) apron feeders and transported directly to the two 18 ft diameter (5.5 m) by 6 ft long (1.825 m)

Cascade mills by parallel 48 in. (1.2 m) wide belts. Because the ore body contains more lenses and masses of quartz than expected, a jaw crusher is to be installed ahead of the semi-autogenous units. The pit-run ore at 8% moisture could not be crushed in a normal crushing circuit to ball mill size feed; hence, crushing-grinding by semi-autogenous methods was adopted. A jaw crusher should solve the problem of oversize, hard boulders.

CARLIN GOLD MINING COMPANY

Carlin, Nevada

Ore type	Hydrothermal gold, siltstones, and limey shales; 143 #/cu ft in place (2891 kg/m^3)
Concentrator capacity	2,500 stpd
Shovel and loader size	5 yd, and 7½ and 10 yd (3.8, 5.7, and 7.6 m^3)
Truck sizes	30, 35, 50, 65, and 75 st
Primary crusher	42" x 48" Pioneer single toggle jaw (1.07 m x 1.22 m)
Connected and consumed horsepower	200 and 165
Capacity of dump pocket	60 st
Method of breaking or removing oversize	Secondary blasting
Open side setting (product size)	4 in. (102 mm)
Crusher feeder	Apron feeder, 54" x 20' 4 3/4" (1.37 m x 6.22 m)
Operating data:	
Shifts operated per week	14
Effective operating hours	6 per shift
Nominal capacity	200 st/hour
Maximum capacity	300 st/hour
Effective availability	95 - 97%
Backing for liners	not applicable
Jaw and cheek plate life	1 year each
Liner materials	Liner plate 400
Capacity of overhead crane	15 tons
Ore bins after crushing plant	Three at 1800 tons each (after secondary crusher)
Direct operating cost per ton	Not available

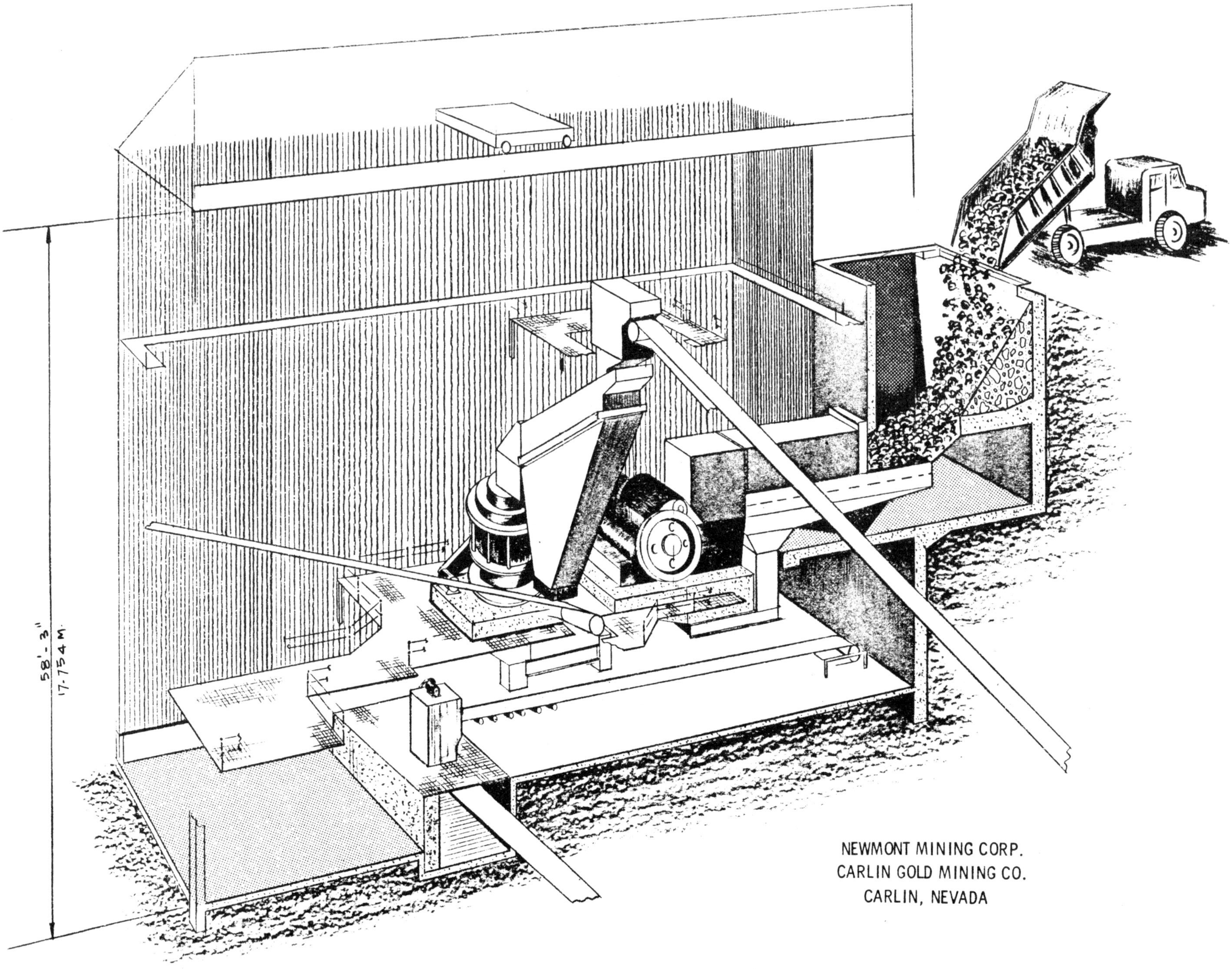

NEWMONT MINING CORP.
CARLIN GOLD MINING CO.
CARLIN, NEVADA

MOLYCORP INCORPORATED
MOUNTAIN PASS OPERATION
MOUNTAIN PASS, CALIFORNIA

The Molycorp operation at Mountain Pass, California, is unique in that it is the only place in the world where bastnasite is mined. The orebody is a metamorphic carbonatite; the mineral of importance is bastnasite, a fluocarbonate containing principally cerium, lanthanum, neodymium, and praseodymium, with small amounts of samarium, gadolinium, and europium. In addition, appreciable quantities of barite, calcite, and strontianite are present.

Molycorp had mined and processed the bastnasite ore since the early 1950s at a rate of about 100 tpd by a unique hot flotation process in a plant originally used for cyanidation of gold ore. The concentrate was primarily used in high quality glass polishing, lighter flints, ceramics, and in specialty steel making. Then in the early 1960s, Molycorp and the Colorado School of Mines Research Foundation developed a solvent extraction process to recover europium from the bulk rare earth concentrates. In 1965 a new SX plant was constructed to produce this material for use as a red phosphor in color television tubes. Because of the demand for this material, a new 500 tpd concentrator was designed and constructed in $9\frac{1}{2}$ months. Then, because of a sharply increased demand for europium, the plant was expanded to 1200 tpd only four months after completion of the original construction.

The crushing section of the new plant was designed similarly to those portable units used in road construction and arrived on the site in ten major pieces. Foundations were concrete piers; and after their completion, the entire crushing plant composed of primary and secondary crushers, conveyors, and a screen was erected in fourteen days by a crew of ten men assisted by a mobile crane. Pit-run ore is loaded into 35-ton trucks by a 7 yd (5.4 cu m) loader, transported to the crushing plant, and dumped into the 65-ton concrete pocket. A 42 in. by 20 ft (1.07 m by 6.1 m) apron feeder feeds ore to the 30 in. by 42 in. (760 mm by 1070 mm) single toggle jaw crusher. Crushed ore is transferred by a scissors conveyor to a 5 ft by 14 ft (1.52 m by 4.27 m) double-deck screen which produces a final 3/4 in. (19 mm) product with the oversize being treated in a $4\frac{1}{4}$ ft (1.3 m) standard cone crusher. Cone crusher product is returned to the first scissors conveyor to join the primary crusher product. Dust collection is by a wet scrubber, and an operator's control room is located to overlook the primary. The entire crushing plant is in the open, and a 30-ton mobile crane is used for maintenance.

Molycorp's two-stage crushing plant is an excellent example of functional design for a small mining and milling operation.

MOLYCORP INCORPORATED
MOUNTAIN PASS OPERATION
Mountain Pass, California

Ore type	Carbonatite - Intrusive calcite/barite/bastnasite
Concentrator capacity	1800 stpd (1633 mtpd)
Shovel size	7 cu yd (5.35 m^3) wheel loader
Truck size	35 st (31.7 mt)
Primary crusher	30" x 42" (0.76 x 1.07 m) Hammermills jaw crusher
Connected and consumed horsepower	125-115 hp (93 - 86 kw)
Capacity of dump pocket	65 st (59 mt)
Grizzly	Not used
Crusher Feeder	42 in. x 20 ft (1.07 m x 6.1 m) apron feeder
Method of breaking or removing oversize	Jackhammer
Open side setting (product size)	3 in. (76.2 mm)
Crusher drive	V-belt
Capacity of crushed ore pocket	none
Crushed ore feeder	Chute
Takeaway belt	30 in. (0.76 m), 400 fpm (122 mpm)
Operating data:	
Shifts operated per week	Ten 8-hour shifts
Effective operating hours	7 per shift
Nominal capacity	120 st (109 mt)
Maximum capacity (peak)	130 st (118 mt)
Effective availability	99%
Backing for liners	None
Jaw plate life	320,000 st (290,000 mt)
Cheek plate life	320,000 st (290,000 mt)
Liner materials in dump pocket and crushed ore pocket	AR Plate
Crushed ore bins, capacity	21,000 st (secondary crushed ore) (19,000 mt)
Capacity of overhead service crane	Mobile crane - 30 st (27.2 mt)
Date of completion	1966

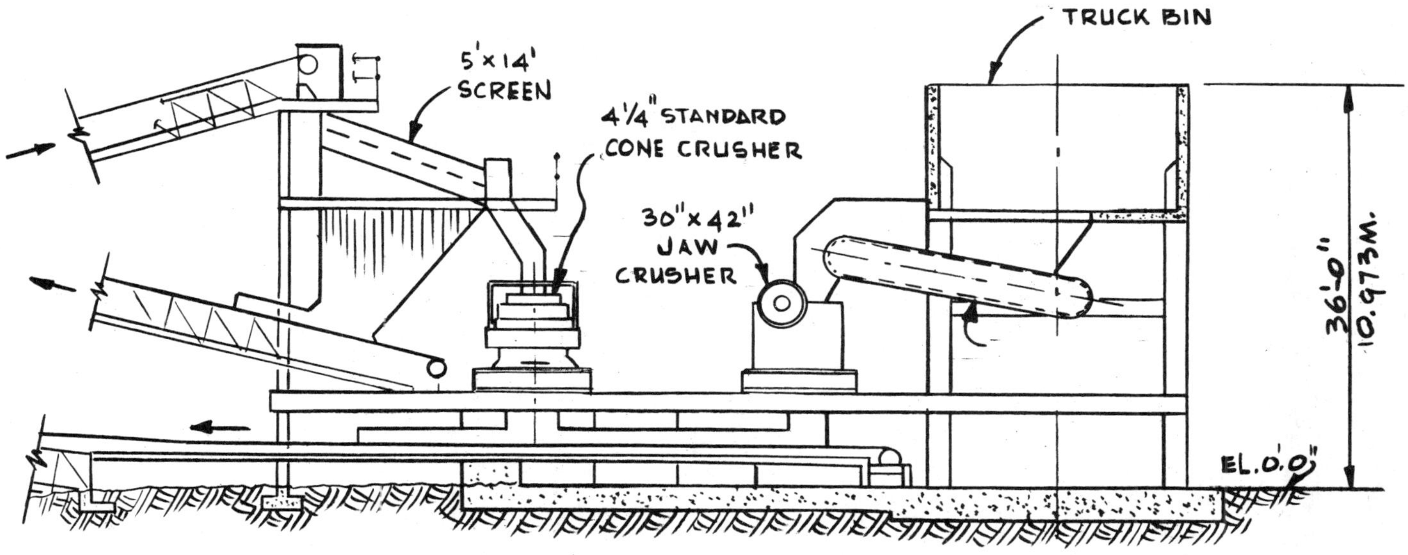

ELEVATION

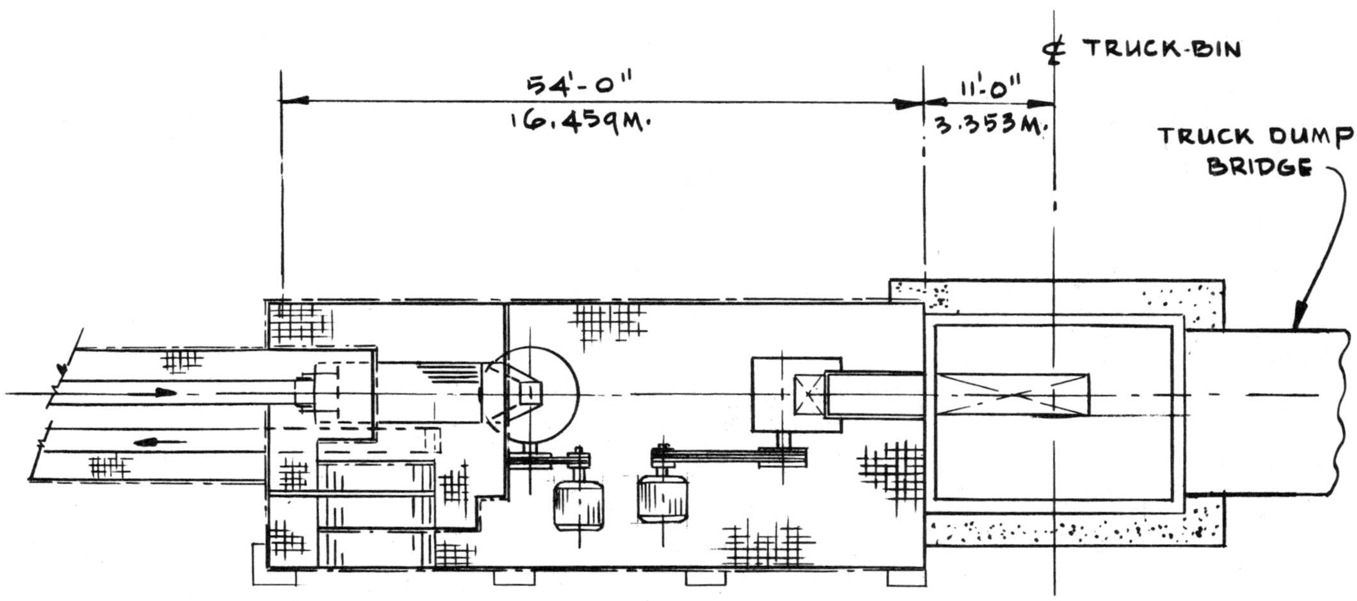

PLAN AT EL. 0'-0"

JAW CRUSHER SIZE	30"x42"/760x1070MM.
CONE CRUSHER SIZE	4'1/4'/1.3M. STANDARD
DUMP POCKET CAP.	65 ST.

MOLYCORP INC.
MOUNTAIN PASS OPERATION
MOUNTAIN PASS, CALIFORNIA

CLIMAX NO. 4

CLIMAX MOLYBDENUM COMPANY
CLIMAX, COLORADO

Molybdenum was first discovered on the Continental Divide in Colorado in 1911; and in 1918 two companies, one of which was the Climax Molybdenum 250 stpd operation, were producing. In 1919 both mines were shut down, but in 1924 the Climax mine reopened and has been producing ever since. Over the years mine production has been increased until today the mill capacity is 48,000 stpd (43,500 mtpd). From 1935 to 1973 all ore was mined underground by block caving with scraper slushing. Since 1973 ore production has been supplemented by open pit mining.

As demand for molybdenum increased, upper levels of the mine became worked-out and new lower levels were opened. In late 1972 production began on the 600 level with the mined ore being primary and secondary crushed in the new No. 4 crushing station. Ore is hauled to the primary in 10 st (9.1 mt) capacity Granby side-dump cars. The trains, carrying 200 st (181 mt), pass through a dump block and discharge directly into a 54 in. by 84 in. Traylor TC gyratory crusher set at an 8 in. (203 mm) open side setting. Since the ore has already been broken to a size which will pass the finger raises in the mining areas, there are only rare oversize boulders to block the crusher opening. Selection of the crusher was based more on its capacity than its receiving opening, as is often the case with gyratories used in open pit operations. Nominal and maximum capacities for this unit are 1250 and 1800 stph (1135 and 1635 mtph). The ore is often wet and sticky, so a mechanical car cleaning device is utilized.

Under the crusher is a 300 st (270 mt) ore pocket equipped with a 60 in. (1.52 m) wide inclined conveyor which delivers the ore to a 2,000 st (1800 mt) surge bin. Near the head pulley of this conveyor is a wood picking station where tramp wood and steel are put into cars for removal.

Two more apron feeders remove the primary crushed ore from the surge bin, and impact belts and conveyor belts transport the ore to the crushing station for screening ahead of two 7 ft (2.13 m) standard crushers. Secondary crushed ore is then conveyed to the surface for tertiary crushing. The entire underground crushing room containing the primary, two secondaries, screens and related equipment is serviced by a 100 st (90 mt) crane with a smaller capacity fast hook. A control room is situated so the operator can observe all the crushing operations. Rotoclone collectors remove dust from the entire operation (including the fumes from the zinc pots) and discharge it to the final product belt. The Rotoclones are vented through drifts and raises to the surface. Shop facilities, electrical gear, and lubrication equipment are located in chambers excavated off the main crushing plant area. An elevator is provided to service the various levels within the crushing plant.

The primary crusher operates 7.4 hours per shift, 21 shifts per week and has an unusually high availability of 93% of scheduled operating time. The 8 hours and 24 hours required for changing out mantles and concaves are a credit to an excellent maintenance organization and are equal to or better than comparable times used in many surface installations. Crushing costs for the primary were $0.06 per short ton for the year 1975.

CLIMAX MOLYBDENUM COMPANY

No. 4 Crusher, 600 Mine Level

Climax, Colorado

Ore type	Disseminated low-grade molybdenum ore 100#/cu ft (1602 kg/m^3)
Concentrator capacity	48,000 stpd (43,500 mtpd)
Train sizes	10 st (9.1 mt) Granby side-dump cars
Primary crusher	Traylor TC 54 in. by 84 in. gyratory
Connected and consumed horsepower	300-200 (226-149 kw)
Capacity of dump pocket	Dump direct to crusher
Method of breaking or removing oversize	All feed goes through crusher
Open side setting (product size)	8 in. (203 mm)
Eccentric throw	2 in. (50.8 mm)
Crusher drive	Direct
Capacity of crushed ore pocket	300 st (272 mt)
Crushed ore feeder	60 in. (1.52 m) apron feeder, variable speed
Speed-up or wear belt	60 in. - 300 fpm (1.52 m - 91.4 mpm)
Takeaway belt	60 in. - 300 fpm (1.52 m - 91.4 mpm)
Operating data:	
Shifts operated per week	21
Effective operating hours	7.4 per shift
Nominal capacity	1250 stph (1134 mtph)
Maximum capacity (peak)	1800 stph (1633 mtph)
Effective availability	93% of scheduled time
Backing for liners	Zinc
Mantle life	3 million st (2.72 million mt)
Concave life	5.4 million st (4.9 million mt)
Time to change mantle	8 hours
Time to change concaves	24 hours
Liner materials in dump pocket pocket and crushed ore pocket	Manganese steel
Capacity of overhead service crane	75 st
Coarse ore stockpile after crusher	2000 st (1800 mt) bin
Direct operating costs (includes labor, power, supplies, maintenance)	\$0.06 per short ton (October 1975)
Concrete	10,768 cu yd (8233 m^3)
Reinforcing bar	524 st (475 mt)
Anchor bolts	7 st (6.35 mt)
Structural steel	250 st (227 mt)
Date of completion	1972

NO. 4 CRUSHER • 600 LEVEL *underground*

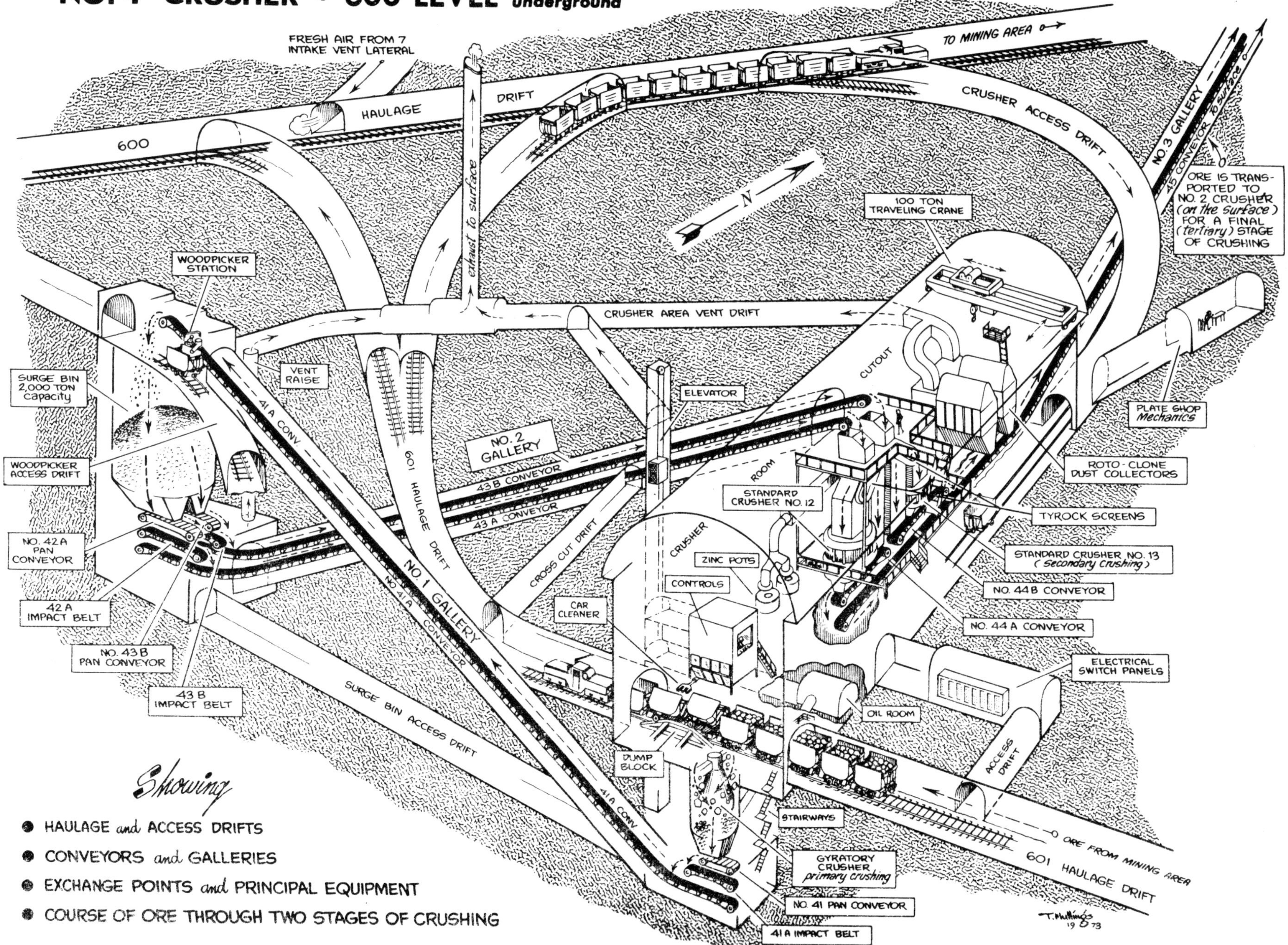

Showing

- HAULAGE *and* ACCESS DRIFTS
- CONVEYORS *and* GALLERIES
- EXCHANGE POINTS *and* PRINCIPAL EQUIPMENT
- COURSE OF ORE THROUGH TWO STAGES OF CRUSHING

CLIMAX MOLYBDENUM COMPANY

GRANDUC OPERATING CO.

GRANDUC OPERATING COMPANY
TIDE LAKE, BRITISH COLUMBIA

Granduc lies near the Alaskan border in British Columbia, about 600 miles (960 km) north of Vancouver. Prospecting in the area must take place in the short summer months of fog and rain because the winter is bitter cold, with snowfall up to 960 inches (24.4 m) accompanied by avalanches. Despite these difficulties, the mineralized area was discovered in the 1920's. By 1956, Newmont Mining Corporation had developed mineable ore reserves estimated at 32,500,000 tons of 1.93% copper. Newmont and ASARCO, in a joint venture, then began development of the ore body, which was partially covered by glaciers.

It was apparent a concentrator could not be built at the mine site where overland access was impossible due to the weather conditions and the terrain. Years of engineering studies of all possible access routes finally dictated the driving of an 11 mi long (17.6 km) tunnel passing under three glaciers and three mountain ranges to the nearest feasible concentrator site at Tide Lake. Even then, a mountain road subject to an average snowfall of 800 in. (20.3 m) had to be built 32 mi (51 km) from the concentrator through Alaska to Stewart, British Columbia, the nearest port. At Stewart, which is located at the head of a fjord 120 mi (192 km) from the Pacific Ocean, a fully equipped marine terminal was built.

The tunnel, the lifeline between mine and mill, was to be driven from both ends; however, in 1965, an avalanche occurred at the Leduc portal, killing 26 crew members and wrecking much of the camp. The tunnel was then driven from one heading adjacent to the concentrator site at Tide Lake, a distance of 53,744 ft (16,381 m) to a point under the ore body. The remote location, type of terrain, and bad weather made supply of crews, equipment, and construction materials difficult; but in spite of these extremely adverse conditions, the tunnel crews set many world records.

Mineralization throughout the ore zone is generally uniform in grade and composition and is primarily chalcopyrite and pyrrhotite associated with pyrite. Mining is mainly by sublevel caving. After blasting, rubber-tired load-and-haul equipment moves from level to level by means of an internal connecting ramp system on a 15% slope. At the gathering haulage level, 15-ton bottom-dump cars deliver the ore to a 600-ton pocket equipped with a Hudson "Rock-Flo" dump.

The bottom of the ore pocket forms the feed chute to the primary crusher. Chain curtains are provided at the draw point to control flow onto a 60 in. by 14 ft (1.52 m by 4.27 m) Hydrastroke feeder, with a rated capacity of 500 tph. The feeder discharges onto a grizzly with 6 in. (152 mm) spacings from which the undersize goes to the takeaway belt. Oversize is the feed to the Traylor 42 in. by 48 in. (1.07 by 1.22 m) jaw crusher. The underground chamber is of sufficient size to accomodate a second crusher if required. The grizzly ahead of an underground crusher is essential for the removal of wet fines. A jaw crusher was a logical choice because of the lower installation head room.

The grizzly fines and crusher product are collected on a 48 in. (1.22 m) transfer belt passing under a tramp iron magnet and then to a 48 in. conveyor inclined at 16°, which transports the ore to either one of the two 3,000 ton ore storage bins. A third bin of 2,000 ton capacity is reserved for waste, which uniquely also has to be crushed, conveyed, and hauled by train to the tunnel portal for disposal.

Ore and waste is drawn from the 3,000-ton rock bins through under-cut arc-loading gates which use the 50-ton railroad cars as volumetric hoppers so the flow ceases when the ore level builds up to the loading chute lip. A train of 20 cars is slowly pulled through the loading track until the cars are full, at which point the gates are closed up upward rotation. Spillage is minimal and all cars are uniformly loaded.

The cars are open hoppers with individual air cylinders arranged for side dumping in either direction. The electric locomotives are capable of pulling twenty 50-ton cars at 40 miles per hour. A round trip to the portal takes slightly more than one hour. At the end of the run, the ore is dumped into a 3,000 ton pocket cut out of rock about 1100 ft (335 m) inside the portal on a spur track.

Layout cost studies were made on locating a crusher underground near the portal or 11 mi (17.6 km) away at the ore body. The ore and waste loading and haulage of run-of-mine size material was the deciding factor in locating the crusher at the mine, even though this meant two hours of lost time in transporting the crusher crews as well as train haulage of all maintenance parts.

Although this crushing plant is installed 11 mi from the tunnel surface exit, the crushing cost of $0.225 per ton attests to an efficient operation.

GRANDUC OPERATING COMPANY

Tide Lake, British Columbia

Ore type	Copper-chalcopyrite with pyrrhotite and pyrite, 2.85 sp gr
Concentrator capacity	7,000 stpd
Loading equipment size	Scooptram ST 5 - 5 yd (3.8 m^3)
Haulage car size	15 st bottom dump
Primary crusher	Traylor 42 in. by 48 in. (1.07 m by 1.22 m)
Connected and consumed horsepower	200 - 165
Crusher feeder	60 in. by 14 ft (1.52 m by 4.27 m) Hydrastroke
Grizzly	
Length and width	10 ft by 5 ft (3.05 m by 1.52 m)
Slope	40 degrees
Spacing of bars	6 in. (152 mm)
Estimated % through grizzly	60%
Method of breaking or removing oversize	Drilling and blasting
Open side setting (product size)	6 in. (152 mm)
Eccentric throw	3 in. (76 mm)
Crusher drive	Belt drive
Capacity of dump pocket	600 st
Capacity of crushed ore pocket	None
Crushed ore feeder	42 in. (1.07 m)
Takeaway belt	48 in. (1.22 m) @ 440 fpm (134 mpm) at 16° to load out bins
Operating data:	
Shifts operated per week	15
Effective operating hours	6 per shift
Nominal capacity	400 stph
Maximum capacity (peak)	500 stph
Effective availability (% of scheduled time)	75%
Backing for liners	Metal to metal
Mantle life	12 months
Concave life	12 months
Liner materials in dump pocket and crushed ore pocket	Not lined
Coarse ore bins	2 @ 3,000; 1 @ 2,000
Direct operating costs (includes labor, power, supplies, maintenance)	$0.2249 per ton (July 1975)
Overall height of installation	43 ft 6 in. (13.25 m) to feeder elevation

GRANDUC OPERATING COMPANY
TIDE LAKE B.C.

L K A B
ALLIANSEN CRUSHER NO. 64
MALMBERGET, SWEDEN

Swedish mining companies are noted for their excellence in underground mining practice, and this classification certainly has to be applied to their design of crushing plants also. All these mines are highly mechanized with either rail or truck haulage, and typical examples of underground crushing plants designed for each of these haulage practices are presented here.

Almost all the primary breakers are of the overhead eccentric, single-toggle jaw crusher type with units as large as 55 by 71 in. (1400 by 1800 mm) being used. In contrast to North America, where jaw crushers are almost always preceded by apron or vibrating feeders, many of the Swedish crushers are choke fed with the dump pocket being located directly above the crusher. This practice results both from excellent knowledge of the ore characteristics and the ability of the overhead eccentric crusher to also act as a feeder. An excellent example of a crushing plant of this design is shown by the sketch of the LKAB Alliansen Crusher No. 64 at Malmberget, Sweden.

Forty-six st (42 mt) trucks dump magnetite ore into a 91.6 cu yd (70 m^3) dump pocket mounted over a 55 by 71 in (1400 by 1800 mm) Morgardshammar overhead eccentric crusher using a 180 hp (132 kw) motor and belt drive. Crushed ore at 10 in. (250 mm) size drops onto an air-operated tipping chute which diverts the ore to either of two 1962 cu yd (1500 m^3) pockets equipped with 3 ft 11 in. by 7 ft 11 in. (1200 by 2000 mm) Schenk vibrating feeders.

Dust is drawn into a wet scrubber and exhaust air is combined with mine exhaust air. Because the crusher must be assembled underground, a 45/5 ton crane is installed over the crushing installation. A crusher above ground such as at Sacaton can easily utilize a smaller crane.

Capacity of this most interesting crushing installation is 1653 st (1500 mtph).

LKAB

Alliansen Crusher No. 64

Malmberget, Sweden

Ore type	Iron ore, magnetite, 2.3 mt/m^3
Shovel size	4.3 cu yd (3.3 m^3)
Truck size	45.4 st (42 mt)
Primary crusher	Morgardshammar 55 by 71 in. (1400 by 1800 mm)
Connected horsepower	180 hp (132 kw)
Capacity of dump pocket	92 cu yd (70 m^3)
Method of breaking or removing oversize	Blasting
Open side setting (product size)	10 in. (250 mm)
Eccentric throw	0.7 in. (18 mm)
Crusher drive	Belt
Capacity of crushed ore pocket	Two at 1500 m^3 each
Crushed ore feeder	3 ft 11 in. by 7 ft 11 in. (1200 by 2000 mm)
Takeaway belt	3 ft 11 in (1200 mm) wide, 295 fpm (90 mpm)
Service crane	45/5 mt
Nominal capacity	1650 stph (1500 mtph)
Jaw plate life	5.5 million st (5 million mt)
Time to change liners	120 man-hours

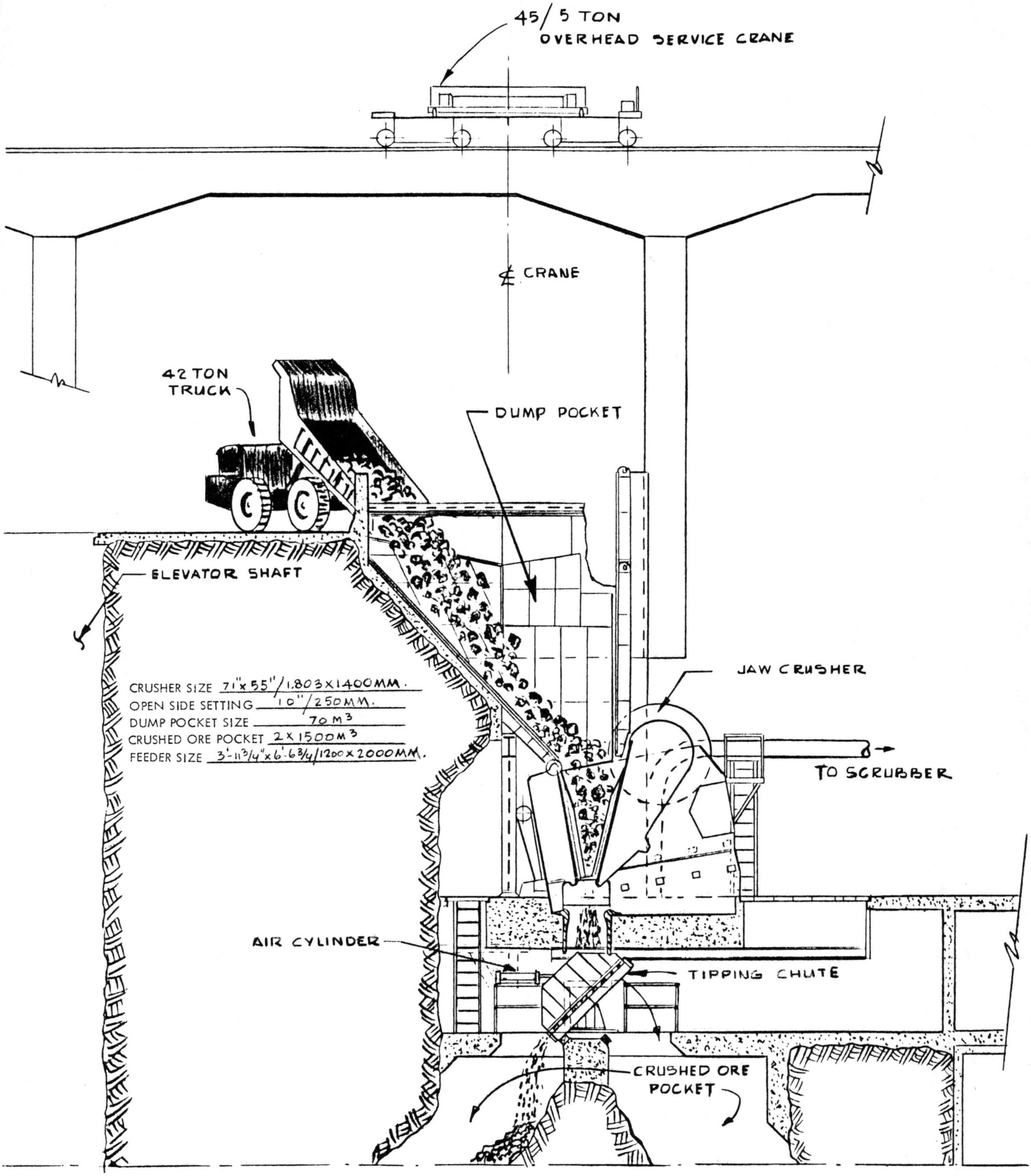

LKAB
MALMBERGET, SWEDEN

L K A B
CRUSHER STATION 51
KIRUNA, SWEDEN

Another excellent Swedish underground crushing plant is that at Crusher Station 51 of LKAB in Kiruna, Sweden. Bottom-dump rail cars pass over a dump scroll to discharge magnetite ore into a 288 cu yd (220 m^3) pocket equipped with a Hewitt Robins E-13 vibrating grizzly feeder. The grizzly section is 59 in. (1500 mm) long and has bar spacings 3 in. (76 mm) wide. Any oversize boulders which will not pass the chute opening to the feeder are blasted. The primary breaker is a Morgardshammar overhead eccentric 55 in. by 71 in. (1400 by 1800 mm) jaw crusher powered by a 200 hp (149 kw) motor. An open side setting of 10.8 to 11.8 in (275 to 300 mm) accounts for the high capacity of this crushing unit (2760 stph or 2500 mtph). This figure compares favorably with the capacity of many gyratory crushers but is typical of the Swedish jaw crushing installations, which are often followed by three more stages of crushing or by autogenous grinding.

Crushed ore from the jaw crusher is combined with the grizzly undersize in the crushed ore pocket and is fed by a second vibrating grizzly feeder to the secondary cone crusher. Both crushers are serviced by 50/5 ton overhead bridge cranes. During maintenance operations air-operated gates shut off the flow of ore from the dump and primary crushed ore pockets.

L K A B

Kiruna Crusher Station KC 51

Kiruna, Sweden

Ore type	Iron ore, magnetite, 2.5 mt/m^3
Primary crusher	Morgardshammar 55 in. by 71 in. (1400 by 1800 mm) jaw crusher
Connected horsepower	200 hp (149 kw)
Capacity of dump pocket	288 cu yd (220 m^3)
Method of breaking or removing oversize	Blasting
Crusher feeder	Hewitt Robins E-13 vibrating grizzly 84 in. by 144 in. (2.1 by 3.66 m)
Open side setting	10.8 in. by 11.8 (275-300 mm)
Eccentric throw	11/16 in. (18 mm)
Crusher drive	Belt
Capacity of crushed ore pocket	78 ½ cu yd (60 m^3)
Crushed ore feeder	Hewitt Robins E-13 vibrating grizzly 84 in. by 144 in. (2.1 by 3.66 m)
Service crane	50-5 mt
Liner materials in dump pocket and crushed ore pocket	Manganese steel
Nominal capacity	2760 stph (2500 mtph)
Jaw plate life	4.96 million st (4.5 million mt)
Time to change liners	75 man-hours

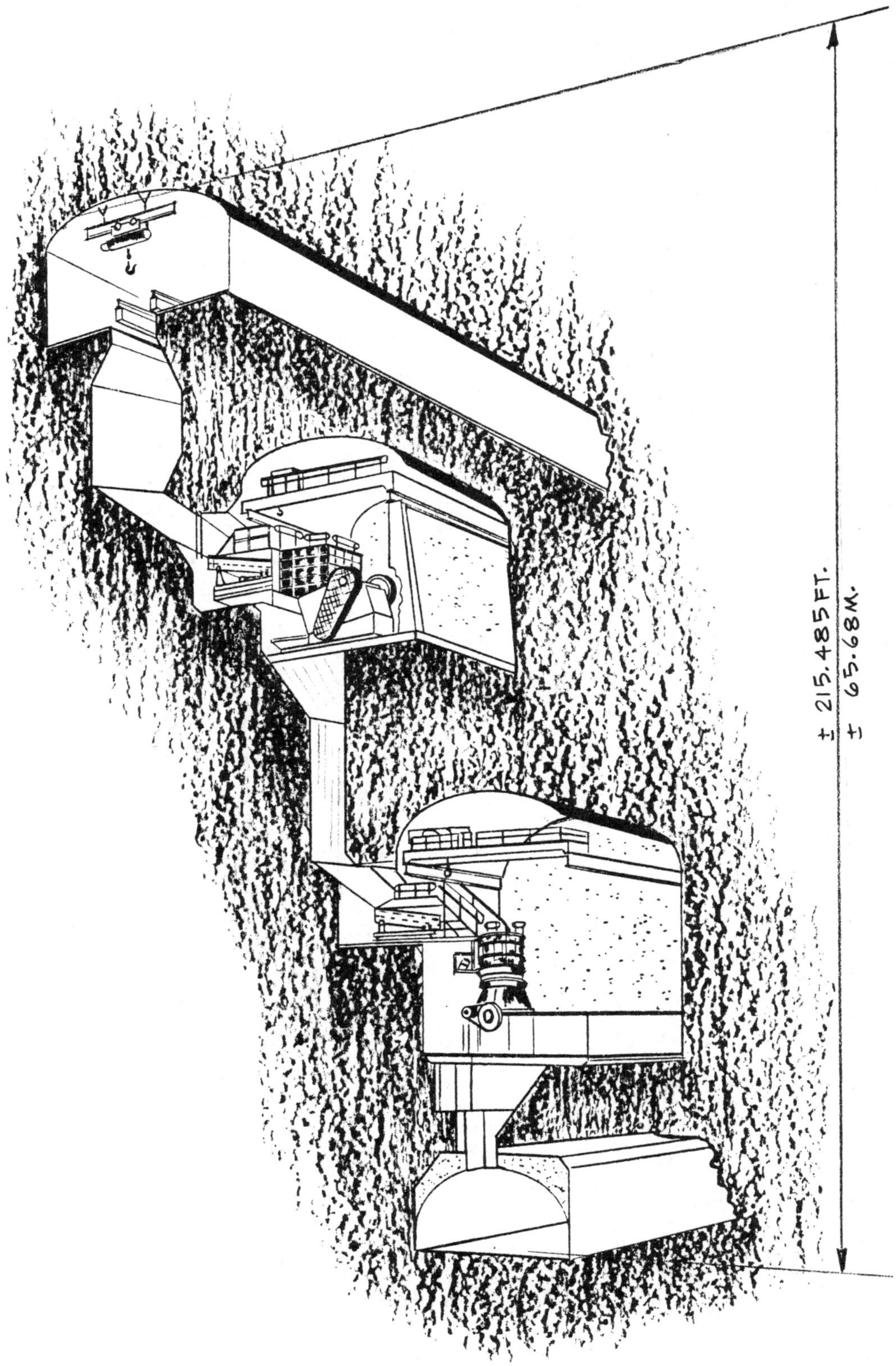

LKAB
CRUSHER STATION 51
KIRUNA, SWEDEN

BIBLIOGRAPHY

"AMAX Dedicates the Henderson Mine-Mill Complex in Colorado," Engineering & Mining Journal. (August 1977).

"Anaconda's New Crusher Complex at Butte, Montana," Skillings Mining Review. (August 26, 1972).

"Anaconda's Twin Buttes Mine in Full Production," Pay Dirt. (April 27, 1970).

Andreas, A., "Mobile Crushing Plant," paper presented at AIME Annual Meeting, Chicago, Illinois, 1973.

Argall, George A., Jr., "Twin Buttes Pit Gets Bigger, 550,000 Tons Moved Out of Pit Each Day," World Mining. (May 1976).

Benson, Norman, "The Granduc Project," Western Miner. Vol. 44, No. 7. (July 1971).

"Bougainville: The New Face in Copper Mining" Engineering & Mining Journal. (February 1973).

"Bougainville Ships Ahead of Plan," World Mining. (October 1972).

Caldwell, A. Blake, "Twin Buttes, A Deep, Low-Grade Copper Producer," Mining Engineering. (April 1970).

Coile, J. J., "In-Pit Crushing and Conveying vs. Truck Haulage," Mining Congress Journal. (January 1974).

"Construction at Duval's Sierrita Mine Will Improve Materials Handling," Engineering & Mining Journal. (November 1976).

Dessau, Max W., and Frank J. Windolph, "Climax Milling Methods," paper presented at AIME Annual Meeting, Chicago, Illinois, 1955.

DeVaney, Fred D., "The Concentration and Pelletizing of Taconite," Mineral Processing. (December 1966).

Dor, A.A., "Recent Trends in Iron Ore Beneficiation and their Effect upon Mill Design and Layout," paper presented at AIME Annual Meeting, St. Louis, Missouri, 1961.

"Dramatic Organization Program Improves Overall Production at Jeffrey Mine," Engineering & Mining Journal. (October 1973).

Faberberg, B., and P. H. Falhstrom, "Crushing and Grinding Practice in Sweden," Mining Engineering. (October 1969).

Gould, Wayne D., "The Pinto Valley Concentrator---Grinding with Large-Diameter Ball Mills," paper presented at SME Fall Meeting, September 1975.

"Granduc Approaches Initial Production Stage," Engineering & Mining Journal. (September 1970).

"Henderson On Stream," Mining Magazine. (January 1970).

Henderson, A.S., E. M. Furness, and F. E. McIntire, "Crushing Practice at Reserve Mining Company," paper presented at AIME Annual Meeting, San Francisco, California, 1959.

Janic, Thomas, "How Engineering, Experience, High-Capacity Equipment Combine to Make Ore at Sierrita," World Mining. (March 1969).

Johnson, N. L., "Rare Earth Concentration at Molybdenum Corporation of America," Deco Trefoil. (August 1968).

"Kennecott Copper Corporation - Ray Mines Division," Pay Dirt. (May 18, 1970).

Knoerr, A. W., "Southern Peru Copper Corporation Starts Production at Toquepala," Engineering & Mining Journal. (April 1960).

Kochanowsky, Boris J., "New Developments in Mobile Crushing Plants are Eyed by Open Pit Operators," Mining Engineering. (September 1968).

Li, Ta M., and Russ Carter, "Pinto Valley Copper Mine - Blue Print for Insured Productivity," Mining Engineering. (June 1975).

LKAB, Giant in Underground Ore Mining," Engineering & Mining Journal. (November 1973).

"Lornex Mine, Managed by Rio Algom Mines, Ltd.," Mining Magazine. (March 1973).

Lutjen, G. P., "Sierrita Makes It with Big Equipment," Engineering & Mining Journal. (August 1970).

McQuiston, F. W., Personal Correspondence, November 1957-1965.

McQuiston, F. W., and R. W. Hernlund, "Newmont's Carlin Gold Project," Mining Congress Journal. (November 1965).

Michaelson, S. D., "Crushing in the Pit," Mining Engineering. (November 1968).

"Mining Program Aims at Growth Opportunities," Engineering & Mining Journal. (August 1974).

Mortensen, Dennis K., "Designing the Sacaton Concentrator," Mining Engineering. (November 1974).

"Palabora," Engineering & Mining Journal. (November 1967).

Parliament, J. Harvey, Private Correspondence, June 1976.

Parliament, J. Harvey, "The Similkameen Project," CIM Transactions. Vol. LXXVI (1973).

"Pinto Valley - Another Major Copper Porphyry Open Pit for Arizona," Mining Magazine. (February 1975).

"Reserve - World's First Large-Scale Pellet Producer," Engineering & Mining Journal. (November 1974).

"Sacaton - Asarco's New Copper Mine in Arizona," Mining Magazine. (February 1975).

"San Manuel History," San Manuel Copper Company, 1940.

Schissler, J. H., "Pueblo Viejo - Rosario's Gold-Silver Producer," Mining Congress Journal. (June 1976).

Shoemaker, R. S., "Inspiration Copper," Private Report, January 16, 1973.

Shoemaker, R. S., "Palabora Figure-8 Crusher," Private Report, April 1975.

Shoemaker, R. S., "Pit Primary Crusher Installations at Twin Buttes," Private Report, January 1974.

Shoemaker, R. S., *et al*, "Field Trip to Visit Ray Mines Primary Crusher," Private Report, March 10, 1970.

"The Similkameen Project," *Western Miner.* (December 1973).

Skillings, D. J., Jr., "Sierrita Copper Project by Duval in Arizona," *Skillings Mining Review*. (January 25, 1969).

Warren, N. F., "Lornex, a Case History," *Western Miner*. (April 1972).

Weiss, Herman, "Mobile Crushing for the Pit," *Engineering & Mining Journal*. (October 1966).

"What's Going on in World Mining - Cuajone," *World Mining.* (September 1976).

"Wood Picking Out of Berkeley Pit Ore," *World Mining.* (September 1972).

Yu, A. T., "Portable Module Systems for Open Pits," *Engineering & Mining Journal.* (June 1967).